Environment in the 21st Century and New Development Patterns

Institute for Global Environmental Strategies

VOLUME 2

Environment in the 21st Century and New Development Patterns

Edited by

K. MATSUSHITA

Institute for Global Environmental Strategies (IGES), Hayama, Kanagawa, Japan

Assistant Editor: Randal Helten

Contributing Authors K. Bezanson, S. Fukukawa, T. Hara, R. Hirono, K. Matsushita, T. Mitsuhashi, A. Morishima, M. Murase, Y. Murota, K. Parikh, H. Srinivas, C. Termorshuizen, H. Tsuchiya, G. Xia, and K. Zhang

Springer-Science+Business Media, B.V.

A C.I.P. Catalogue record for this book is available from the Library of Congress.

ISBN 978-90-481-5617-7 ISBN 978-94-015-9670-1 (eBook)
DOI 10.1007/978-94-015-9670-1

Printed on acid-free paper

This is a translation of the original Japanese work *21 Seiko no Kankyo to Shinhatten Patan,* Chuo Hoki Publishing Company, IGES 1999.

Photo credits: K. Matsushita

Contents

Preface — 1

A. MORISHIMA

Book Editor and Chapter Co-Authors — 5

Introductory Chapter: Asian Environment and New Development Patterns — 9

K. MATSUSHITA

Introduction
Need for New Development Patterns
Recent Environmental Trends in Asia and the Pacific Region
The Financial Crisis and the Environment
Peace Making through Environmental Cooperation
Towards Rio +10
Structure and Outline of This Book
Recycling-oriented, Decentralized and Networking Society
Development Patterns in Asia: Diverse and Plural Value Systems
New Concepts of Development
Concluding Remarks

Section I: New Development Patterns in the 21st Century Society

Chapter 1. Material Cycles and the Formation of an Environmentally Sound Market Economy — 27

T. MITSUHASHI

Japan's Material Cycle
The Zero Emissions Approach
Transition to an Economy Based on Stocks
Creating an Environmentally Sound Market Economy

Chapter 2. Energy and New Development Patterns — 47
H. TSUCHIYA
Introduction
Global Energy Future
Improving Energy Efficiency
Decentralizing Energy Supplies
DES and the Economy
From Energy Hunting to Energy Cultivating Civilization
Conclusion

Chapter 3. Energy and the New Economy — 69
H. TSUCHIYA
Introduction
The New Economy
Aspects of the New Economy
Effects of the New Economy on Resources and Energy
Energy Systems in the New Economy
Conclusion

Chapter 4. Impacts of the Information Technology Revolution: Economic Challenges of New Development Patterns — 87
Y. MUROTA
Introduction
What is the IT Revolution?
Social Impacts of the IT Revolution
The IT Revolution and New "Development"
From Industrial to Information Technology Societies

Chapter 5. The Role of Information Disclosure in Corporate Governance: The Case of the Chemical Industry — 115
C. TERMORSHUIZEN
Introduction
The Goals of Corporate Governance
The Role of Information in Corporate Governance
Mechanism of Corporate Governance
The Role of the Media
Strengthening Information Disclosure Mechanisms
Conclusion

Chapter 6. Sustainable Water Strategies for New Development Patterns — 147
M. MURASE
Introduction
A Change in Thinking
The New Paradigm
What Must Be Done

Chapter 7. Toward Sustainable Agriculture and Development —— 163
T. HARA
Background
Steps Toward Sustainable Agriculture
Free Trade and Environmental Destruction
Sustainable Agriculture Providing a Basis for the New Agricultural Basic Law
Summary

Section II: New Development Patterns and Regionality

Chapter 8. New Development Patterns in Asia —— 201
S. FUKUKAWA
Implications of Global Environment
Dynamism in Asian Countries
Asian Values
Contribution of Asian Sense of Values to Restoration of the Global Environment
Framework of Measures Required to Pursue Sustainable Development
Collaboration within APEC
Globalization and the Importance of the Environment

Chapter 9. New Development Patterns in India —— 215
H. SRINIVAS
Introduction
Development Challenges for India
Current Development Patterns
Emerging Patterns
Summary

Chapter 10. New Development Patterns and Innovation in China's Environmental Policy —— 239
ZHANG K. and XIA G.
New Development Patterns and Environmental Policy
Fundamental Characteristics of China's Environmental Policies
Evaluation of China's Environmental Policy
The Direction of China's Environmental Policy Innovation
Concluding Remarks

Section III: New Development Patterns and International Development

Chapter 11. International Development at the Dawn of the 21st Century — 257

K. BEZANSON

IGES in Context as a New International Organization
Development - Past and Present
Rethinking Development and the Language of Development
Globalization: Order and Disorder
The Future of Development Cooperation

Chapter 12. Enjoy it by Giving it up: Toward Sustainable Development Patterns — 271

K. PARIKH

Need for Development
Present Development Patterns
Changing Patterns
New Development Patterns
Concluding Comments

Chapter 13. New Development Patterns in Asia - Securing Economic Growth, Social Justice, Healthy Environment, Sustainable Human Development and Good Governance in the 21st Century — 291

R. HIRONO

Introduction
Long-term Vision for Asian Countries for the 21st Century: Great Challenges and Opportunities
Mutually Reinforcing Reforms of Economic, Social, Environmental and Governance Policies
Developing National Strategies in Asian Countries Compatible with New Development Patterns

Afterword — 301

K. MATSUSHITA

Preface

Akio Morishima
Chair of the Board of Directors of IGES (Institute for Global Environmental Strategies)

Here at the dawn of the twenty-first century, humanity faces the challenges left us from the last one hundred years, an era when advanced industrialized cities and countries glittered in the light of material-based civilization, awesomely visible even from space.

One facet of the remarkable progress of industrialization is the unprecedented consumption of natural resources and consequent environmental destruction wrought on a global scale. We have changed the very chemical composition of the atmosphere and patterns of weather–some fundamental conditions of nature on our planet.

Without bold changes in direction the future is likely to bring painfully realized predictions of accelerated resource depletion and irreversible damage to the environment. Meanwhile, away from the prosperity enjoyed in the richest countries, much of the two-thirds of the world's population eking out an existence in developing countries still suffers from hunger and poverty.

Years ago, in preparation for the United Nation's 'Earth Summit'–the 1992 United Nation's Conference on Environment and Development in Brazil–the World Commission on Environment and Development (Brundtland Commission) released a report entitled *Our Common Future* in 1987. The report said that progress was necessary in developing countries so that every person could be ensured the basic material needs in order to live a life with dignity, while stressing that this development should not cause unlimited consumption of resources and destruction of the environment. The concept of "sustainable development," was introduced–development that fulfills the needs of the present generation without reducing the capacity to meet the needs of future generations. This concept challenges developed countries to reflect on their uncontrolled growth in the past, and also hints at a new path of development for developing countries.

Japan achieved rapid economic growth based on its economic policies represented by the income doubling plan of the 1960s. The other side of this phenomenon was serious cases of industrial pollution that resulted in tragic health impacts, such as in Minamata (ethyl mercury poisoning from factory effluent) and Yokkaichi (asthma caused by sulfur oxide emissions from factory smoke). Based on this experience Japan introduced strict pollution regulations and developed leading-edge pollution control technologies. Though the image as a polluted country was not exactly an enviable one, through its experience of overcoming pollution Japan did learn lessons that may be helpful in addressing global environmental problems.

Starting in 1989 the government took on global environmental issues when it boosted Japan's official development assistance to developing countries in the environmental field as a way to contribute to world peace, and not to militarism. This approach was adopted partly because the international community at the time looked to Japan as a major economic power to use this power for international cooperation. But that was not all; Japan also thought that its experiences in overcoming industrial pollution–and the technologies and social systems that made this possible–would contribute to problem-solving in Asia's newly industrializing countries.

After the Earth Summit, the term "sustainable development" became almost a household word. However, since that time the developed countries have not changed their social infrastructures of mass consumption and disposal, and developing countries appear to be hurrying along the same path of industrialization.

The key word "sustainable" was given to us, but answers to key some questions are not yet forthcoming. Exactly what conditions are sustainable? And who must do what in order to make society sustainable?

In 1995 the Committee on the Global Environment of the 21st Century, a private advisory committee, proposed to Japan's then Prime Minister Murayama that a policy research institute should be established as soon as possible to study ways to answer the questions of how to realize a sustainable society. The government acted on this proposal, and as a part of Japan's international contribution to address global environmental problems, decided to establish a research institute to conduct international, interdisciplinary, practical research on these issues. After three years of preparation, the Institute for Global Environmental Strategies was established in March 1998. Now into its third year, IGES is conducting research projects in six areas: climate change, urban environmental management, forest conservation, environmental education, environmental governance, and new development patterns.

This book is one outcome of the activities of the New Development Patterns Project. As mentioned, the world today is still groping for answers to questions about which development paths will allow humanity to realize sustainable societies. Each project of IGES searches in its own field for approaches to environmental protection that are sustainable and compatible with development and progress.

The New Development Patterns Project did not start with a specific, preconceived concept for patterns of sustainable development. Instead, the goal of its ongoing research is to get closer to a concrete image of these patterns. At the project launch, discussions were held involving experts in various fields. The discussions included somewhat of a re-consideration of the development patterns of the twentieth century and new thoughts about what kind of advances and developments can be envisioned for the twenty-first century.

This book is one of the fruits of those efforts and, we hope, a guide to light the way for future research. The views and opinions of the thinkers who contributed to this book contain a wealth of ideas. I hope that the publication of this book will deepen public interest in sustainable societies and stimulate far-reaching discussions.

IGES intends to continue publishing key outcomes of this project that carry social significance. In that light I would like to express my sincere appreciation to Kluwer Academic Publishers for their understanding of IGES' objectives, and for undertaking the publishing of this series.

Book Editor and Chapter Co-Authors

(in alphabetical order)

Keith A. Bezanson

Dr. Bezanson is the Director of Institute of Development Studies, University of Sussex. He received his PhD in 1972 from Stanford University. A citizen of Canada, Dr. Bezanson has been President of Canada's International Development Research Centre (IDRC) from 1991-1997 and previous to that was Vice-President of the Canadian International Devlopment Agency (CIDA). He was also Canadian Ambassador to Peru and Bolivia in 1985-1988 and Administrative Manager of Inter-American Development Bank in 1988-1991. He is the author of numerous publications on international development.

Shinji Fukukawa

Mr. Fukukawa is the Chief Executive Officer, Dentsu Institute for Human Studies, Dentsu Inc. He graduated from the University of Tokyo in 1955, and joined the Ministry of International Trade and Industry (MITI). In 1986 he was the administrative vice-minister of the ministry. After retirement, he was an advisor to MITI/RI (Research Institute) and Nomura Research Institute, the vice-president and the vice-chairman of Kobe Steel, Ltd. before taking the present position.

Takeshi Hara

Prof. Hara has been teaching at Waseda University Graduate School, Japan since March 1998. Prof. Hara graduated in Law from Waseda University in 1961, and worked in Mainichi Newspaper Company for thirty-seven years, where he held various positions including editor of the Science and Environmental News Department, senior staff writer and editorial writer. He currently serves as a Director of Nature Conservation Society of Japan and The Association of National Trusts in Japan.

Ryokichi Hirono

Prof. Hirono is Professor Emeritus at Seikei University, Japan. Having graduated from the University of Chicago in 1959, he worked as research economist at Japan Management Association and then joined Seikei University in 1961 as Assistant Professor in its Faculty of Politics and Economics, moving up the ladder to become

Professor in its reorganized Faculty of Economics in 1970. He also worked as Chief, Development Planning Division, UN/ESCAP during 1974-76 and as Assistant Administrator of UNDP in charge of Policy and Evaluation during 1986-89, with the rank of Assistant Secretary General of of United Nations.

Kazuo Matsuhsita

Mr. Matsushita is the Acting Vice-President of the Institute of Global Environmental Strategy (IGES). He received his MA in Political Economy from Johns Hopkins University in 1976. Mr. Matsushita has been working for the Environment Agency of Japan (EAJ) since 1972, and his experience includes Administrator in Environment Directorate of OECD, Senior Program Officer at UNCED, Director of Air Pollution Control Division and Global Environment Conservation Division of EAJ, and Director General of Department of the Japan Fund for Global Environment, Japan Environment Corporation.

Tadahiro Mitsuhashi

Prof. Mitsuhashi has been appointed as professor at Chiba University of Commerce since 2000. He graduated from Keio University, Tokyo, in 1964, and worked for Nihon Keizai Shimbun Inc. as a journalist for thirty-six years. His experience includes correspondent in Brussels in 1976-1979, London bureau chief in 1985-1988, senior chief editorial writer in 1992 - March 2000.

Akio Morishima

Prof. Morishima is the Chair of Board of Directors of the Institute for Global Environmental Strategies (IGES). He graduated from the University of Tokyo in 1958, and was at Nagoya University for more than thirty-five years, as associate professor, and professor and Dean at the School of Law, and as the Dean of Graduate School of International Development. In 1997 he became a professor at Sophia University. He has been involved in the Central Environment Council of Japan as the Chairman of Policy Planning Committee since 1993, and has been appointed as the President of the Council in 2000.

Makoto Murase

Dr. Murase is working at Department of Environmental Protection, Sumida Ward Office, Tokyo. Sumida City, which has been promoting rainwater utilization projects, received Local Initiatives Awards as the Finalist for Excellence in Freshwater Management from International Council for Local Initiatives Awards in June 29, 2000. Dr. Murase received his PhD from Toho University in 1996 and his MSc in 1976 from Chiba University. He is also Director of Urban Affairs, International Rainwater Catchment Systems association, Secretary General of People for Rainwater Utilization, and a part-time lecturer at the Department of Pharmaceutical Science in Toho University.

Yasuhiro Murota

Mr. Murota is the representative director of Shonan Econometrics, developing software and building programs. He graduated from the University of Tokyo in 1965, and he was a senior research fellow at the Japan Center for Economic Research, and assistant professor at Saitama University before taking the present position.

Kirit S. Parikh

Prof. Parikh is the Director of Indira Gandhi Institute of Development Research since 1986. He received his Sc.D. in Engineering in 1962 and M.S. in Economics and Engineering in 1963 from Massachusetts Institute of Technology (MIT). After working at MIT as a research assistant and associate, he was appointed as Professor of Economics at Indian Statistical Institute. Prof. Parikh also worked with International Institute for Applied Systems Analysis (IIASA), where he was a leader of Food and Agriculture program in 1980-1986.

Hari Srinivas

Dr. Srinivas is a Programme Associate at the United Nations University in Tokyo, Japan. He received his MSc from Asian Institute of Technology, Thailand, in 1991, and his PhD from Tokyo Institute of Technology, Japan, in 1996. He has worked as a Research Associate at AIT and at the Global Environment Information Center (Tokyo), before working as an Assistant Professor at Tokyo Institute of Technology during 1997-1999.

Cindy Termorshuizen

Ms. Termorshuizen is currently in Canada. She received her MA in Political Science from Carleton University, Canada, in 1994. Ms. Termorshuizen lived in Taiwan between 1993-1994, where she worked as an intern in environmental sector of Canadian Trade Office, and in Japan between 1994-2000. In Japan, she was involved in the Institute for Global Environmental Strategies as an Intern and Researcher, and worked as Foreign Affairs Aide to Akiko Domoto, a member of the House of Councilors of Diet of Japan, and also the President of GLOBE international, and Vice-President of IUCN.

Haruki Tsuchiya

Dr. Tsuchiya is the President of the Research Institute of Systems Technology. He received his PhD in Engineering from the University of Tokyo in 1970. Dr. Tsuchiya established the Research Institute of Systems Technology in 1979, and has been working in the field of energy analysis and computer sciences.

Xia Guang

Dr. Xia Guang is the Deputy Director and Associate Professor of Policy Research Center for Environment and Economy of State Environmental Protection Administration of China. He is also a member of the Experts Committee of the Sino-Japan Cooperation Project for Environmental Demonstration Cities. Dr. Xia graduated from China Renmin University with Ph.D.in Economics and has been working in the field of theoretical policy issues related to environment and economy, especially environmental policies in China.

Zhang Kun

Mr. Zhang Kun is the Director-General of Sino-Japan Friendship Environmental Protection Center in Beijing. He graduated from Dongbei University in 1968, and then worked as an engineer in Junchuan Nonferrous Metal Company, served as the chairman of city council of People's Congress of Junchang city, Gansu Province. He also was the Derector-General and senior engineer of Environmental Protection Bureau of Gansu Province.

Introductory Chapter

ASIAN ENVIRONMENT AND NEW DEVELOPMENT PATTERNS

KAZUO MATSUSHITA
IGES (Institute for Global Environmental Strategies)

1. INTRODUCTION

Agenda 21, adopted at the Earth Summit held at Rio de Janeiro in 1992, presented a comprehensive action plan for a sustainable society in the twenty-first century. In particular it sought to eliminate poverty as well as to facilitate a paradigm shift, from the present development path which is based on the wasteful use of resources, to a more sustainable one (UN, 1992). In order to realize the paradigm shift, the methods for putting the agenda into practice must be clarified, and analytical research is needed to suggest strategic policies and to bring them about.

More than eight years have passed since Agenda 21 was adopted. From a global perspective, have we advanced toward a paradigm shift and sustainable development advanced?

This kind of generalized assessment is certainly not easy to make, but *Global Environmental Outlook, 1997* made the following assessment: "...despite this progress on several fronts, from a global perspective the environment has continued to degrade during the past decade, deeply imbedded in the socio-economic fabric of nations in all regions." (UNEP, 1997)

In order to curb damage to the environment, and to remove structural problems, developed and developing countries must together establish new development patterns (NDP). These patterns should encourage both consideration for the environment and sustainable development.

The new development patterns that IGES is seeking have a global perspective, but their focus is particularly on the Asian region. This focus is partly due to IGES' own original reasons for coming into existence. It also reflects the fact that Asian

K. Matsushita (ed.), Environment in the 21st Century and New Development Patterns, 9–24.

countries will be critical for the solutions to global environmental problems in the twenty-first century, because of their huge populations and continuing rapid economic growth-countries such as China, India, ASEAN countries, Korea, Taiwan and others. The development paths that these countries choose to follow from now on will greatly influence their impact on the environment. The ability of many of these countries to enjoy the benefits of their development will be undermined by the rapid changes they are undergoing and the economic and social distortions and weaknesses which these changes will bring. In contrast to the dominant style of economic society-a one-way path of over-consumption of resources-the world is in dire need of another path that turns toward new development patterns. In order to create this new path, joint efforts of the countries involved and research institutes are necessary.

2. NEED FOR NEW DEVELOPMENT PATTERNS

The twentieth century was a period of rapid economic growth and expansion. The world economy grew by about five times between 1950 and 1998, from approximately 5 trillion dollars in 1950 to almost 29 trillion in 1998. The world population grew nearly four times from the beginning to the end of the century from 1.6 billion to today's 6 billion. About 60 percent of today's global population

Figure 1. Projection of World Population.

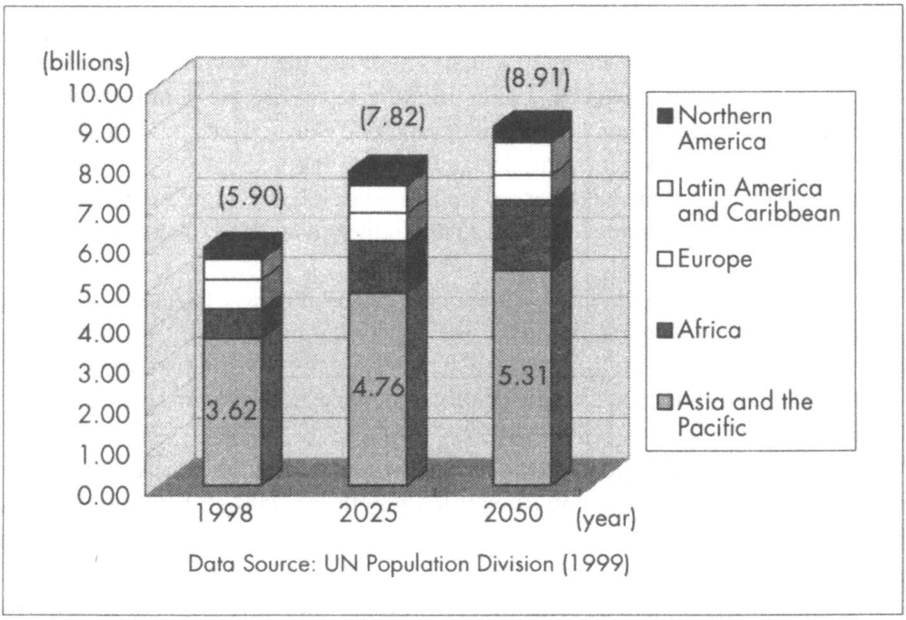

Figure 1 has been prepared by the IGES-NDP Project based on the estimates and projections published by United Nations Population Division (1998 revision).

or 3.6 billion people, live in Asia (Figure 1).

Economic growth after World War II was made possible through the rapid expansion of the use of natural resources and energy, and growth in economic and social systems. These had the effect of accelerating patterns mass production, mass consumption and mass disposal. However, the expansion of industrial activities and their supporting economic and social systems spread unevenly over the globe, concentrating in Western Europe, North America and Japan. Many regions of the world remained unable to enjoy the fruits of economic growth. Even today, the gap between developed countries and developing countries is widening. Moreover, it is becoming increasingly apparent that current development patterns followed by industrialized countries are causing serious environmental problems and that they are neither ecologically nor socially sustainable. The limits of the ecological carrying capacity of the Earth are becoming apparent.

In recent years, many Asian developing countries, which have suffered from poverty for decades, are experiencing high rates of economic growth by introducing material-intensive patterns and consumption patterns based on throw-away products. The accelerating trend of globalization and the fast development of information technology magnify on a global scale the problems associated with current development patterns.

Society as a whole is facing critical challenges today. How can we transform existing economic and social systems into more environmentally benign and sustainable ones in order to realize ecological sustainability and social equity? The answer is that we must identify and put into practice new development patterns for the twenty-first century, in both developed and developing countries.

3. RECENT ENVIRONMENTAL TRENDS IN ASIA AND THE PACIFIC REGION

The Asian landscape is marked by immense diversity, from the northern steppes to deserts and tropical rain forests. Asia has virtually every type of ecosystem found on Earth. The environmental problems the region faces, consequently, are equally varied: soil erosion, deforestation, urbanization and industrialization, lack of safe drinking water, and hygienic concerns raised by waste and sewage treatment. In this sense, the Asian environment has many characteristics representative of the global environment.

Population and poverty issues remain high on the policy agenda in this area. The United Nations announced that the world population passed the 6 billion mark on October 12, 1999. Although growth rates are likely to decline, the global human population is still increasing. The Asia and the Pacific region is densely populated; the 3.6 billion people it holds (1998) live on only 30 percent of the Earth's entire land area. The population in the region is expected to reach 4.8 billion by the year 2025 and 5.3 billion by the year 2050.[1]

The annual average growth of per capita GDP in Asia and the Pacific was 3.09 percent (1975-95), far above the average global growth rate of 1.17 percent (UNEP, 1999). Most countries in the region experienced rapid economic development and industrialization from the mid-1980s to the mid-1990s, and some benefit was delivered to regional inhabitants. However, it is also true that drastic social changes increased inequality in the distribution of wealth. Thus, a recent report of UNEP entitled GEO2000 points out that poverty still remains a significant problem in the region, particularly in South Asia.[2]

Another worrisome concern is the adverse impact of urbanization. Cities around Asia (excluding Japan and the "four dragons" of Hong Kong, Singapore, South Korea, and Taiwan) are seeing their populations grow by 5 to 6 percent annually, and new megacities with more than 10 million people are emerging throughout the region.

According to the United Nations Economic and Social Commission for Asia and the Pacific (ESCAP), a large proportion of urban dwellers live in slums or as squatters; as of 1993 the proportion was over 50 percent in Indonesia, above 45 percent in Bangladesh, nearly 40 percent in India, and close to 30 percent in the Philippines (Figure 2). The availability of fresh water is another serious problem. Both urban slums and water problems are growing worse due to rapid population growth and industrialization.

Figure 2. Proportion of Urban Population in Slums/Squatters.

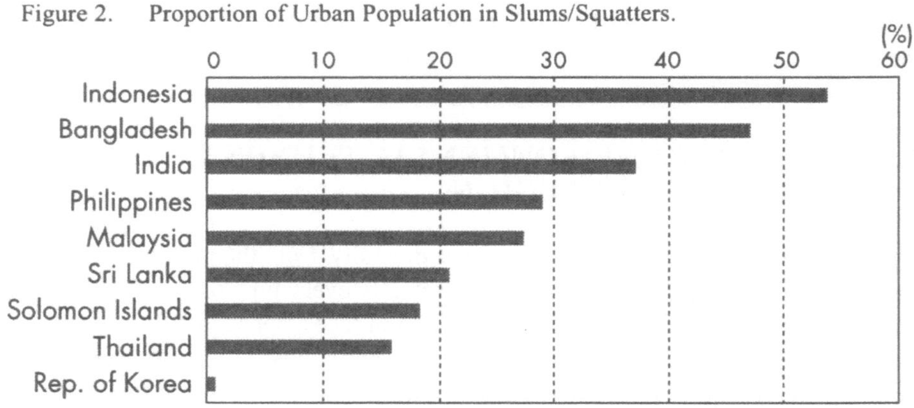

Source: ESCAP (1993). The State of Urbanization in Asia and the Pacific 1993. UN, New York

Asian countries are forced to address these environmental issues-which the industrial nations of the West were able to grapple with over the two centuries since the Industrial Revolution, and which Japan has tackled since the Meiji Restoration-while at the same time battling poverty.

Environmental headaches will only grow if current trends continue. As energy consumption increases, emissions of sulfur oxides and carbon dioxide are expected to rise, and 15 percent of what was forest in 1990 could become farmlands or grassy plains by 2025.

Primary energy consumption in East, Southeast and South Asia has grown rapidly. It increased from 675 million tons of oil equivalent in 1971 to 2.23 billion tons in 1996 (Energy Data and Modeling Center, 1999). The growth rate between 1990 and 1996 was as high as 4.5 percent per year.

The growth of energy consumption has caused environmental degradation, especially air pollution. The air quality in Asia's cities is among the worst in the world. Levels of air pollution in many cities substantially exceed the international standards for air quality set by the World Health Organization (WHO). Ten of Asia's 11 mega-cities exceed WHO guidelines for particulate matter by a factor of at least three, four exceed acceptable lead levels, and three exceed acceptable ozone and sulfur dioxide levels (ADB, 1997).

The Asian-Pacific Integrated Model (AIM) developed by Japan's National Institute for Environmental Studies (NIES) forecasts that energy consumption in this region, which now accounts for just over 20 percent of world energy consumption, will increase to more than 30 percent (Government of Japan, 1999).

Figure 3. Projection of CO_2 Emissions.

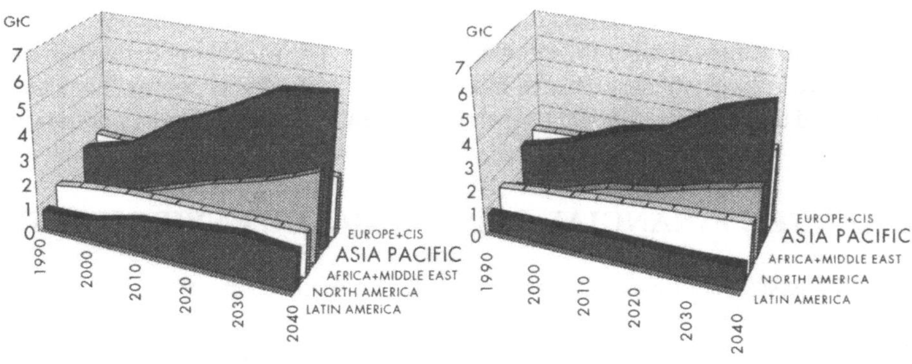

(a) Projection of CO_2 emission under the high growth scenario

(b) Projection of CO_2 emission under the low growth scenario

Source: AIM model

The High growth scenario is based on assumptions in which globalization of market economies continues; innovations take place in the field of education, technologies, and social systems; and rapid and high economic growth is achieved worldwide, particularly in developing countries. The population growth rate is projected to be low. The Low growth scenario is based on assumptions in which international economic cooperation is limited and recovery from economic crises or recessions do not progress in short-term; economic development mainly relies on domestic resources, in terms of both natural and technological resources; and low economic growth. The population is projected to grow very rapidly.

As a result, large quantities of air polluting substances such as sulfur dioxide, and greenhouse gases such as carbon dioxide, are projected to be released into the atmosphere (Figure 3).

A number of recent disasters are symbolic of encroaching environmental deterioration. The June 1998 flooding of the Yangtze River in China caused 3,600 deaths and affected 230 million people. Some 48 million hectares of farmlands were damaged, and total damage from the flood reached some 260 billion yuan (around U.S.$32 billion).

China has frequently been struck by devastating floods over its long history. It is important to note that, Chinese top government leaders, including Premier Zhu Rongji, believe that flood damage was exacerbated by deforestation along the middle and upper reaches of the Yangtze River; they thus regard floods not just as natural disasters but as "ecological and environmental problems" induced by human activities. The Chinese government has since banned the felling of virgin forests along the middle and upper stretches of the river.

Whereas the Yangtze is prone to flooding, the Yellow River suffers from a chronic water shortage. In 1997, for example, the mouth of the river was dry for a period of seven months. Farms, factories, and residents that draw on the downstream sections of the Yellow River become embroiled in heated disputes over water rights.

Haze caused by forest and peat fires in Indonesia is another big issue. Fires that raged in the second half of 1997 affected Malaysia and Singapore, causing air pollution and disrupting traffic. The government in Jakarta and NGOs concur that the primary culprit is the expansion of oil palm plantations; a secondary factor is the industrial afforestation of fast-growing pulpwood. The fires exposed the intimate links between forest management in Indonesia and the global economy. They also highlighted the need for regional cooperation to deal with transboundary pollution.

4. THE FINANCIAL CRISIS AND THE ENVIRONMENT

The 1997 East Asian financial crisis triggered a dramatic economic contraction in Asia and the Pacific. The average growth rate plummeted from 4.0 percent in 1997 to minus 6.9 percent in 1998 in the economies of Southeast Asia, and from 6.0 percent to minus 1.4 percent during the same period in the newly industrialized economies (ADB, 1999). Currency devaluation resulted in substantial consumer price inflation in 1998, averaging 21.0 percent in Southeast Asia and 13.0 percent in South Asia (ibid.). Current accounts across the region showed some improvement, but this was caused largely by foreign capital outflows and import reductions.

The recovery from the sharp regional recession seemed to begin in late 1998. Several major economic indicators, such as real growth rates, foreign exchange reserves, inflation rates, and equity markets, show that 1999 was a year of rebuilding for economies of the whole region. Undervalued currencies, together with growing demand from the United States and Europe, where the economies were still robust, increased Asia's exports by 1 percent, while the region recorded the strongest import contraction in volume and value (Review Publishing Company, 2000). Intra-regional trade rose by 2 percent, after a 15 percent decline in 1998 (ibid.). Despite the recovery being regarded as uneven and fragile, the World Bank expected East Asia as likely to become the fastest-growing developing country region in 2000 (World Bank, 1999a).

The East Asian financial crisis affected virtually every aspect of life in the region, including the environment. According to one report, the financial and environmental crises had common roots: the pursuit of rapid growth without consideration of proper safeguards, policies and controls (World Bank, 1999b). The report also highlighted the following features of short and long-term impacts of the financial crisis on environment.

In the short-term, it was reported that the economic contraction resulted in improvements in air pollution due to a reduction of traffic congestion in metropolitan areas and emissions of pollutants from factories. However, the positive effects of the crisis on urban air quality were transitory.

In the longer term, the crisis has thrown the region's delicate ecological balance out of kilter, as devastated rural communities have been forced to exploit the natural resources around them-forests, fish and shellfish, and minerals-to secure their own survival, an option also encouraged by national governments desperate for export earnings.

In addition, industrial and municipal treatment facilities were forced to cut back operations and untreated wastewater, solid waste discharge, and illegal dumping increased. In some countries, budget cuts placed environmental programs in jeopardy.

Until the East Asian financial crisis occurred, it was evident that the benefits of economic growth were being directed into environmental improvements. However, as a result of a reversal of this trend after the crisis, the risk of environmental neglect may be growing.

In one respect, though, the task of drawing up a road map to recovery presents an excellent opportunity-not only for Southeast Asia, but for every country that is involved in the region-to reconfirm the goal of sustainable development that harmonizes environmental concerns with development and lifestyles. This can set in motion initiatives to achieve more ecologically friendly and healthier modes of development. Such initiatives could spark an economic recovery fueled by environment-related infrastructure improvements and strategic environmental investments.

In order to lay the groundwork for long-term, sustainable development, government spending to boost demand should be directed at building eco-friendly facilities and enhancing energy efficiency. For instance, water supplies systems that recycle rainwater or incorporate water-saving devices, sewage and waste water treatment plants that match local needs, energy efficient public transportation systems, renewable energy, and material recycling systems.

Tokyo's Sumida Ward and 24 other local public entities offer subsidies for water-conserving devices, such as equipment to store, utilize and facilitate the ground permeation of rainwater, and to recycle rain water. Earth friendly technologies like solar heating, solar power generation, wind power, biomass, and cogeneration are spreading at the community level, and with subsidies or other incentives, they could begin to meet a much higher share of energy needs.

Strategic investments in equipment and infrastructure to promote environmental conservation will improve environmental efficiency for industry. They will also ensure the competitiveness of national industries in a global market that will demand compliance with stiffer environmental requirements, including those to reduce greenhouse gas emissions.

Many developing countries do not have adequate funding, technologies and know-how to deal with global environmental issues like climate change; their priorities are securing safe drinking water, addressing local air pollution, treating household waste, and dealing with shortage of natural resources like firewood. The international community, therefore, should direct its capital and technology to these countries to help promote ecologically sustainable development.

One possible approach is to address regional and global issues at the same time. This can enhance the efficiency of environmental policies and trigger new investment and technology flows.

Measures to improve energy efficiency, for instance, will help alleviate air pollution and carbon dioxide emissions. Industrialized countries that have made commitments under the Kyoto Protocol (agreed upon in December 1997 at the Third Session of the Conference of the Parties to the United Nations Framework Convention on Climate Change) to reduce greenhouse gas emissions, and the developing countries coping with local pollution should coordinate their activities more closely. New mechanisms to channel funds and technology should be established. Specifically, the Clean Development Mechanism now being considered under the Kyoto Protocol is promising: it could encourage industrialized nations to make energy-saving investments in developing countries that contribute to a cleaner environment and reduce carbon dioxide emissions at the same time.

5. PEACE MAKING THROUGH ENVIRONMENTAL COOPERATION

How can environmental cooperation contribute to peace and security? The term "security" has traditionally referred to measures adopted by national governments to counter military threats; it included the forging defense of alliances, stepping up of economic cooperation, maintaining diplomatic neutrality and other means.

Subsequently it has been applied more broadly to mean safeguarding the lives of citizens by ensuring economic growth and supplies of energy and food. This concept of "comprehensive security" has been augmented further by the idea of "human security."

Toward the end of Cold War in the late 1980s, arguments promoting the linking of environmental and security issues emerged in the United States. Jessica T. Mathews, then vice president of the World Resources Institute, argued that the deterioration of global environmental resources caused by problems, such as tropical deforestation, desertification, and shortages of fresh water, could threaten regional security (Mathews, 1989). She further contended that environmental issues should be added as important factors to discussions of national and global security. Advocates of "environmental security" subsequently embraced the argument that environmental degradation could trigger regional conflicts.

Typical examples of this are conflicts over water resources in the Middle East and the emergence of environmental refugees and ethnic clashes triggered or worsened by ecological deterioration in Africa. Regional resources that could cause such conflicts, therefore, must be managed more closely with the agreement and understanding of the concerned parties.

When considering Asian environmental cooperation, this notion of environment and security should be advanced a step further: the notion of regional cooperation for peace. Collaboration on environmental projects can foster trust between the parties involved. While military showdowns are fundamentally zero-sum in nature (i.e., one side wins while the other loses), environmental cooperation can be a plus-sum game (both sides win).

Regional environmental cooperation is an essential element of effective responses to trans-boundary pollution. Its purpose is to tackle international environmental problems through collaboration among countries at risk. In the Asia-Pacific region, several regional cooperation forums have been established, mainly at the sub-regional level.

The Acid Deposition Monitoring Network in East Asia (EANET), for instance, is a promising, ongoing example of regional cooperation to address trans-boundary air pollution. An interim report of RAINS-ASIA, a joint World Bank-Asian Development Bank project, predicts that acid deposits over an extensive area in Asia will exceed ecologically sustainable levels and surpass levels in industrial countries. To counter this trend, the Environment Agency of Japan proposed the creation of EANET, which has the central objective of helping create a common understanding of the issue of acid rain in East Asia by gathering and analyzing scientific data. The first expert meeting was held in 1993, and the initiative is now in its preparatory stage, involving ten countries in the region. It is to be formally established at its second intergovernmental meeting in the year 2000 (Nii, 1999). The network is expected to provide a model of regional cooperation for environmental management in Asia and the Pacific.

The environment is an area, moreover, where cooperation in East Asia is possible despite its disparate military and political orientations and levels of economic development. The Wild Bird Society of Japan, for instance, has been conducting research on the migratory patterns of the black-faced spoonbill, a species of heron threatened with extinction, with the cooperation of groups in China, Hong Kong, North Korea, South Korea, Taiwan, and Vietnam. There are estimated to be only 600 black-faced spoonbills surviving, based on winter sightings in Japan (around Hakata Bay and the Ariake Sea off the shores of northern Kyushu). To protect this species, its breeding grounds, migratory routes, and winter homes must be located quickly and protected from further deterioration. This protection is especially important in areas where the birds breed. The spoonbills were caught and tagged while wintering, and their routes have been monitored by satellite with technical assistance from Nippon Telegraph and Telephone. After three years of research, their breeding grounds are becoming clear: the southeast coast of Dalian in China and the shores near the North-South Korean border. It is hoped that further cross-border cooperation will lead to more concrete steps to protect this species.

The fact that other cooperative mechanisms also began to flourish during the last decade shows that cooperation with neighboring countries has strengthened in the Asia-Pacific region. Another example is the North-East Asian Sub-regional Program of Environmental Cooperation (NEASPEC) which has engaged six countries since 1993 in projects such as reducing pollution from coal-fired power plants, environmental monitoring, data collection and analysis.[3] The South Pacific Regional Environmental Program (SPREP), consisting of 26 countries and territories, has continued to evolve since its establishment in 1982. SPREP officially became an autonomous regional organization in 1995, and has implemented various projects, such as bio-diversity and natural resource conservation, climate change mitigation, integrated coastal management, and education about waste management.

Another exemplary initiative is an annual forum of Chinese, Japanese, and South Korean environment ministers, launched in January 1999 at the suggestion of the South Korean government. At the February 2000 meeting held in Beijing, an agreement was reached for Japan and South Korea to assist in an afforestation project in inland China. Other priority areas for East Asian cooperation include marine conservation in the northwest Pacific, global warming countermeasures, improvement of urban environments, and forest conservation.

Further development of these regional cooperation forums will bring about enhanced trust among countries in Asia and the Pacific. Trust is a fundamental contributor to international stability and peace. Countries should promote efforts to strengthen existing programmes, to create new regional collaborative schemes, and to build upon the networks of those international frameworks.

6. TOWARD RIO + 10

In 2002 a UN-sponsored conference is scheduled to commemorate the tenth anniversary of the 1992 Earth Summit in Rio de Janeiro and to undertake a comprehensive review of Agenda 21. "Rio+10," as it is being called, will bring together political leaders from around the world to discuss directions to change the existing patterns of development toward new twenty-first century-type sustainable patterns.

The meeting will survey the state of the environment and is expected to produce agreement on new directions for society and concrete policy package to ensure ecological sustainability for the global community. As Asia is expected to have a large bearing on sustainable development in the twenty-first century, contributions from this region will be a significant factor in determining the success of the conference.

At the Rio Summit, Prime Minister Kiichi Miyazawa pledged a leading role for Japan in the preservation of the global environment and committed to boost environment-related official development assistance by around nine hundred billion to one trillion yen over a period of five years. Actual disbursements since then greatly exceeded this commitment. Later, at the June 1997 special session of the UN General Assembly to appraise post-Rio implementation, Prime Minister Ryutaro Hashimoto outlined Japan's "Initiative for Sustainable Development Toward the 21st Century." Its guiding principle was to promote "human security," "ownership" (self-help efforts of developing countries), and environmental cooperation toward sustainable development. As specific areas of action, it included measures to fight pollution, address global warming, promote natural environment conservation, secure fresh water resources, and enhance environmental awareness and strategic research.

Prior to the Rio+10 meeting, Japan must review its domestic measures to build an environmentally sustainable society and objectively assess its track record of foreign aid in this field. By pursuing intellectual and policy dialogue with Asian and other countries, it should strive to play a leading role in the Rio+10 process through the formulation and implementation of sustainable development policies. It goes without saying that Japanese society itself must pursue efforts to recycle more and place less of a burden on the environment.

7. STRUCTURE AND OUTLINE OF THIS BOOK

While bearing in mind the problems that have been described so far, experts from Japan and abroad, including IGES directors and trustees, were requested to write papers on various issues relevant to new development patterns.

Amongst these there are papers which (1) focus on a special field, (2) are written from the point of view of a specific country or region, and (3) develop awareness of problems that have general application to new development patterns. This book comprises three sections that follow the same categorization.

Section 1 includes the following:

Chapter 1 "Environmentally Conscious Market Economy Formation and Materials Cycle" by Tadahiro Mitsuhashi
Chapter 2 "New Development Patterns and Energy" by Haruki Tsuchiya
Chapter 3 "New Economy and Energy" by Haruki Tsuchiya
Chapter 4 "The Economic Challenges of New Development Patterns: Impact of the Information Technology Revolution" by Yasuhiro Murota
Chapter 5 "The Role of Information Disclosure in Corporate Governance - The Case of the Chemical Industry" by Cindy Termorshuizen
Chapter 6 "A Sustainable Water Strategy for New Development Patterns" by Makoto Murase
Chapter 7 "Movement Toward Sustainable Agriculture and Development" by Takeshi Hara

Section 2 includes the following:

Chapter 8 "New Development Patterns in Asia" by Shinji Fukukawa
Chapter 9 "New Development Patterns in India" by Hari Srinivas
Chapter 10 "New Development Patterns and the Innovation in China's Environmental Policy" by Zhang Kun and Xia Guang

Section 3 includes the following:

Chapter 11 "International Development on the Eve of the 21st Century" by Keith Bezanson
Chapter 12 "Enjoy it by Giving it up - Towards Sustainable Development Patterns" by Kirit Parikh
Chapter 13 "Development and Good Governance for the 21st Century" by Ryokichi Hirono

8. RECYCLING-ORIENTED, DECENTRALIZED AND NETWORKING SOCIETY

Section 1 introduces various issues, which are relevant to the realization of new development patterns. Each of the authors was requested to discuss separate themes and wrote independently. Naturally, the proposals and emphases in each paper reveal common elements in understanding of the problems. Below is a brief introduction of the papers.

Mitsuhashi's paper starts with the recognition that the Earth has resource and environmental limits and conditions are deteriorating. It proposes a "zero-emissions" approach and the building of an environmentally-conscious market economy in order to shift from the current throw-away economy based on mass production, mass consumption and mass disposal to an economic system which is more recycling oriented and causes less burden on the environment. The paper then emphasizes, in developed countries in particular, a change from flow-based to stock-based economies. The image of the stock based economy contains the following key words: stable economic growth, appropriate production, long-lasting products, service-oriented manufacturing industry (focussing on the provision of service rather than provision of physical products), domestic demand oriented, and decentralization.

Tsuchiya's first paper proposes increasing the efficiency of energy use, a move towards decentralized energy systems, a shift from being an energy "hunting" civilization to an energy "cultivating" civilization, and the concept of a new economy with highly efficient use of energy and resources.

Tsuchiya's second paper considers the possible energy-saving benefits of the New Economy, based on the revolution in information technology. The use of electronic mail, teleconferencing, telecommuting and other changes in the way we live and work have the potential to reduce the amount of energy needed in society.

Murota's paper discusses the information technology revolution and global environmental problems. It suggests that, based on the IT revolution, industry will have the possibility of moving away from manufacturing, that communications will substitute traffic, and that through technological "leap-frogging" by developing countries, the load on the environment may be reduced.

Termorshuizen's paper argues that information disclosure programs and rules instituted in the mechanism of corporate governance work as driving forces to promote innovation, competitiveness, and environmental care. Taking the chemical industry as an example, the conditions to ensure such processes were discussed.

Murase's paper focuses on strategies relating to water resources and proposes moving away from dependence on river sources to independence, and from destruction of regional water cycles to symbiosis. It also discusses the idea of cyclical capacity, shifting from demand control to supply control, and the use of networks.

Hara's paper is based on a number of case studies. He stresses the concept of sustainable agriculture from the viewpoint of "material cycles," the coexistence of human activity with natural ecological systems, the joint participation of producers and consumers, and agriculture as the pillar to sustain regional communities.

The points of view that emerge from these papers promote self-sufficiency, autonomous approaches based on regional empowerment and networking, and societies that break away from materialism and manufacturing.

9. DEVELOPMENT PATTERNS IN ASIA: DIVERSE AND PLURAL VALUE SYSTEMS

In Section 2, new development patterns in Asia were discussed in three different contexts.

Fukukawa's paper, based on analysis of the history of Asian development and value systems, proposes directions of efforts to restore the global environment, frameworks for sustainable development and possible international cooperation.

Srinivas' paper evaluates the impacts of economic liberalization, which took place in India during the early 1990s. It also summarizes the development patterns in India and presents future projections.

Zhang and Xia's paper reviews recent developments in Chinese environmental policies. It then considers the direction of policy innovations based on the concept of long-term sustainability of environment and resources.

These papers provide an Asian perspective and plural values to our discussion.

10. NEW CONCEPTS OF DEVELOPMENT

Section 3 is composed of three papers on new development patterns, contributed by three experts on international development from two industrialized and one developing country.

Bezanson's paper looks back the past and the present of international development. It proposes that we fundamentally reconsider the concept of "development" to accommodate changing international frameworks under the current trend of globalization. Under a new concept of development, it underlines the importance of region-specific diverse values and the carrying capacity of the environment.

Parikh's paper confirms that current development patterns are neither ecologically, socially nor politically sustainable. Based on that recognition, it then tries to identify desirable patterns of development, how to realize them, and the kinds of research required.

Hirono's paper lists areas that economic development and integration in the 1980s failed to address, such as equitable income distribution, elimination of poverty, and environmental sustainability. It then stresses the need for economic growth, which is compatible with social equity, sustainable development and good governance. In order to attain such objectives, it calls for drastic policy and institutional reforms as well as promotion of investment in human capital, science, and technology.

Currently, globalization is spreading and affecting societies around the world. The common theme of the authors of this section concentrates on how to realize sustainability under this reality. How should we identify new patterns of development which are different from existing ones? What kind of development patterns should be followed by Asian countries while combating poverty? The papers have identified the issues and orientations to approach the problems. However, further research will be required to make the ideas more concrete.

11. CONCLUDING REMARKS

This book is a compilation of contributions of experts on various issues which comprise important ingredients to analyze new development patterns, particularly in Asia and the Pacific. Certainly there is a long way to go to arrive at the final goal of the research on new development patterns. This book is but a preliminary study on this important subject. We hope it will stimulate further work in this important area of research and that it will provide insights for its readers.

REFERENCES

ADB (Asian Development Bank). 1997. *Emerging Asia.*

ADB. 1999. *Asian Development Outlook 1999.*

Energy Data and Modeling Center, *The 1999 Handbook of Energy and Economic Statistics.*

Government of Japan. 1999. *White Paper on Environment 1999.*

Institute for Global Environmental Strategies. 1998. *1998 Top News on Environment in Asia.* 1998.

Matsushita, K. 1999. *Peace Making through Environmental Cooperation in the Asia-Pacific*, Outline of Presentation at Energy, Security, Environment in Northeast Asia (ESENA) Conference, 1999.

Matthews, J.T. 1989. *Redefining Security*, Foreign Affairs 68.

Nii, M. 1999. Presentation Draft for ECO ASIA '99.

Review Publishing Company. 2000. *Far Eastern Economic Review Yearbook Asia 2000*, Hong Kong: Review Publishing Company Ltd.

United Nations. 1992. Agenda 21: Programme of Action for Sustainable Development, Report of the United Nations Conference on Environment and Development, Rio de Janeiro, 3-14 June 1992 (A/CONF.151/26/Rev.1).

United Nations Development Programme. 1997. *Human Development Report 1997*, Oxford University Press, New York, United States, and Oxford, United Kingdom.

United Nations Economic and Social Commission for Asia and the Pacific and the Asian Development Bank. 1995. *State of the Environment in Asia and the Pacific.*

United Nations Environmental Program. 1997. *Global Environmental Outlook 1997*.

1999. *Global Environmental Outlook 2000*. http://www.unep.org/geo2000/english/index.htm

United Nations Population Division. 1999. *World Population 1998*, http://www.undp.org/popin/wdtrends/p98/fp98.htm

World Bank. 1999a. *Annual Report 1999*.

World Bank. 1999b. *Environmental Implications of the Economic Crisis and Adjustment in East Asia*, Discussion Paper Series No. 01.

ENDNOTES

1 The figures cited for 2025 and 2050 are based on the medium fertility variant of the United Nations Population Division estimates and projections as revised in 1998 (UN Population Division, 1999).

2 Some 75 percent of the world's poor live in Asia. There are 515 million people living under the income poverty line ($1 PPP a day) in South Asia and 446 million people under the line in East Asia, Southeast Asia, and the Pacific (UNEP, 1999; UNDP, 1997).

3 For further information on NEASPEC refer to http://unescap.org/enrd/environ/neaspec.htm

Section I

New Development Patterns in the 21st Century Society

Chapter 1

MATERIAL CYCLES AND THE FORMATION OF AN ENVIRONMENTALLY SOUND MARKET ECONOMY

TADAHIRO MITSUHASHI
Chiba University of Commerce

Humanity must transform its economic systems in which materials flow 'one-way' through mass-production, mass-consumption and mass-waste. We must introduce resource-recycling economic systems that put the lowest possible burden on the environment. Economic development must be built on the recognition that our planet has limits and the environment is being degraded. The Zero Emissions approach is a powerful means to achieve these goals. In order to utilize this approach effectively, an environmentally sound market economy is essential.

1. JAPAN'S MATERIAL CYCLE

1.1 Annual Flow of Materials -- 2.2 billion tonnes

When seen from the point of view of the material flows, economic activity is nothing more than the process of taking resources and using them to make products that are then distributed for consumption. Waste is the product of this process, appearing in the production stage and again in the distribution and consumption stages.

According to data from the Environment Agency of Japan, 1.95 billion tonnes of resources are used each year (1997 figures) in the course of the nation's economic activity. Of this amount, 1.18 billion tonnes originate in Japan, 700 million are imported in the form of raw materials and 70 million are imported in the form of manufactured products (Figure 1-1).

K. Matsushita (ed.), Environment in the 21st Century and New Development Patterns, 27–46.

Figure 1-1. Material Balance in Japan

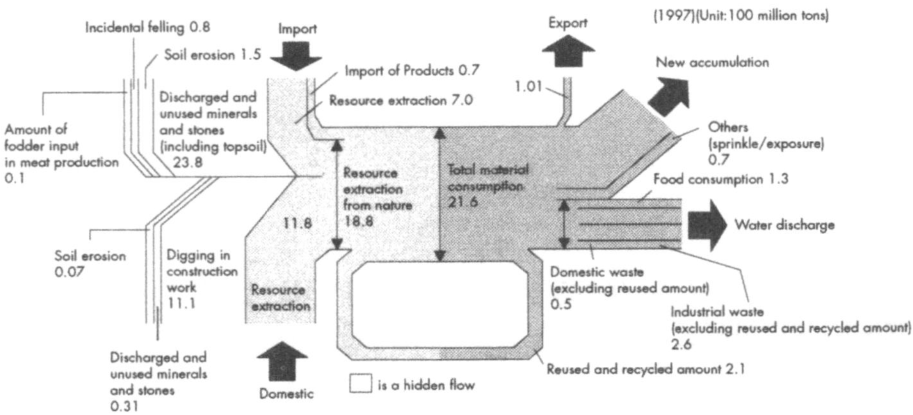

Note: the total amount from the producing country is larger than total amount of material used due to the water content of raw material.
Source: calculated by the Environment Agency of Japan with various statistics.

Resources Used (billion tonnes) 1.95	=	Domestic Resources 1.18	+	Imported Raw Materials 0.7	+	Imported Products 0.07

In addition to the above amounts are 210 million tonnes of resources that are 'recycled' and then re-used. This brings the total resources used to 2.16 billion tonnes. This resource is mainly used for manufacturing products. Part of the products are exported, and there are wasted produced in the process.

The total 2.16 billion tonnes of used resources end up with 310 million tonnes of gross waste after subtracting the amount for manufactured products, exports and energy consumptions. Of this waste, 50 million tonnes are general waste, excluding those recycled, and 260 million tonnes are industrial waste.

There is already a national shortage of waste disposal plants for this general and industrial waste (Table 1-1, Figure 1-2).

Table 1-1. Residual volume and period of final desposal sites for industrial waste (as of April 1995).

	Amount of waste needed to be disposed (million tonnes)	Residual volume of the sites (million m^3)	Period that the sites can accommodate the wastes (years)
the Tokyo metropolitan area	21.28 (23.6)	20.07 (17.72)	0.9 (0.8)
the Kinki area	11.59 (12.6)	40.4 (44.09)	3.5 (3.5)
Nationally	80 (84)	208 (210.7)	2.6 (2.5)

Note:

(1) The Tokyo metropolitan area includes the following prefectures and metropolis: Ibaragi, Tochigi, Gunma, Saitama, Chiba, Tokyo and Kanagawa. The Kinki include following prefectures: Mie, Shiga, Kyoto, Osaka, Hyogo, Nara and Wakayama.

(2) The amount of watste needed to be disposed in the Tokyo metropolitan area and the Kinki area were calculated by:
80 million tonnes X 26.6 % for the Tokyo metropolitan area, and
80 million tonnes X 14.5 % for the Kinki area.
The percentages are taken from the data in 1994.

(3) Residual periods of the sites were calculated as residual volume / amount needed to be deposited.

(4) Figures in parentheses are for 1994.

Figure 1-2. Changes in residual volumes and periods of final disposal sites for general wastes. (from "Quality of the Environment of Japan 1998", Environment Agency of Japan)

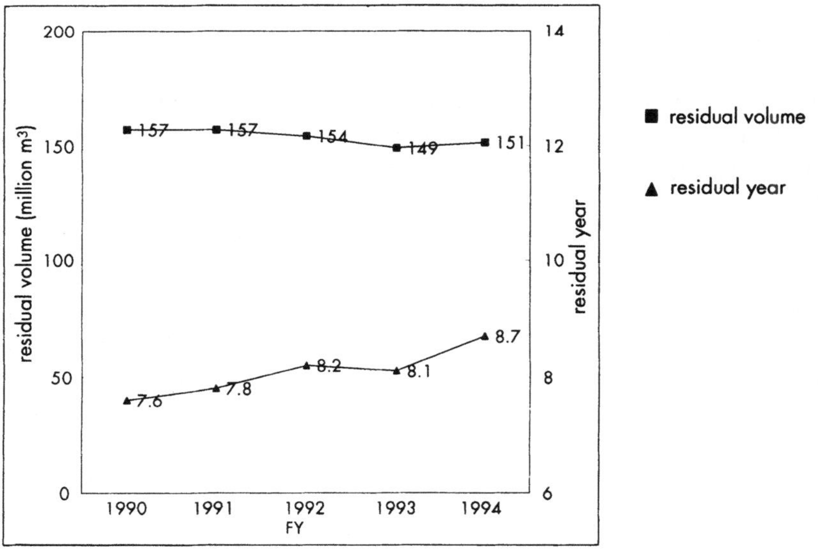

Figure 1-3. The proportion of minerals that have already been mined.

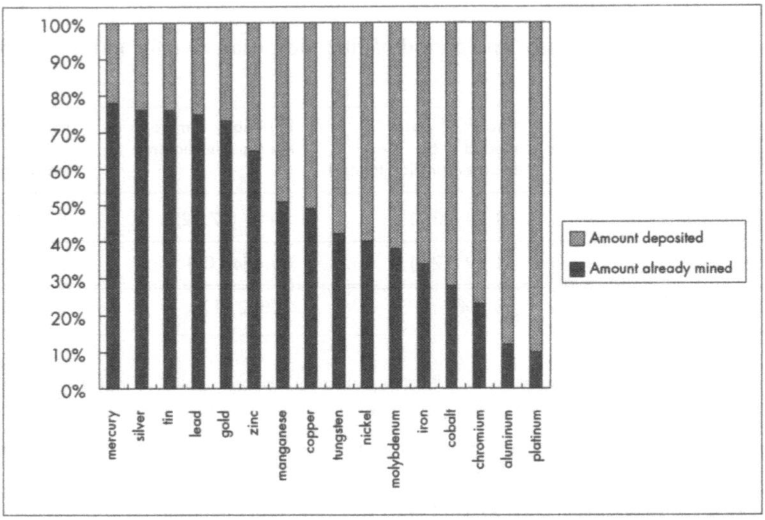

Source: *"Shigen Keizaigaku no susume"* (in Japanese), Takashi Nishiyama, Chuko Shinsho, 1993

1.2 Resource Depletion

Natural resources are being rapidly depleted worldwide as a result of the rapid economic development of the past one hundred years. More than 70 percent of estimated reserves of metallic resources like mercury, silver, tin, lead and gold, have already been excavated or used; a mere 30 percent remains underground. Fifty percent of estimated reserves of zinc, manganese and copper and 30 percent of the tungsten, nickel and iron have already been excavated, although these minerals are still perceived as being unlimited. Oil reserves are quickly being depleted, and are forecasted to be exhausted in 40 years if consumption continues at the current rate. The effective use of waste will therefore take on added importance in the twenty-first century as a means to prevent the depletion of natural resources (Figure 1-3).

Experts predict that economic development and deforestation caused by growing populations will worsen world water shortages and that in the twenty-first century, conflicts over water resources will replace conflicts over oil. Humanity must address the need to avoid further degradation of the global environment, and the need to wisely use natural resources necessary for economic development such as metals and oil, in order to avoid exhausting them. In addition, we must develop alternative renewable resources, and, while striving to avoid the generation of waste, seek ways to recover resources from the waste that is inevitably produced.

2. THE ZERO EMISSIONS APPROACH

2.1 Construction of an Industrial Ecosystem

In this new century humanity must change from its 'one-way' economic system based on mass-production, mass-consumption and mass-waste, to another one that places the smallest possible burden on the environment. The Zero Emissions approach is a powerful tool to realize this ideal.

Emissions may be defined as waste or discharge, so Zero Emissions means, literally, "zero waste". The approach has recently been used in a larger context to refer to economies, societies, regions, or companies that do not discharge waste. The theory behind the Zero Emissions approach comes directly from natural ecosystems. It is said that there are no components of the natural world that fail to serve a purpose. Every thing or body inevitably has some use for something or somebody else. In the food chain, herbivorous animals eat vegetation and carnivorous animals feed on them, but when they die they return to the soil and are decomposed with the help of fungi and bacteria and become the nourishment of plants. The natural ecosystem is sustained by this repeating cycle. If it were possible to apply similar approaches to a society's economic activity, a zero-waste society could be achieved. In short, trying to construct an economic 'ecosystem' comparable to natural ecosystems is the Zero Emissions approach.

2.2 The Five Revolutions

To build a Zero Emissions society, five revolutions are needed.

2.2.1 Design Revolution

First is a revolution in product planning and design. Raw materials are necessary to make products. If all the raw materials used to make the product were integrated into it, there would be no waste. Input (raw materials) equals output (products). Until now, large quantities of raw materials have been wasted per unit of product produced, because resources were thought to be unlimited. Today we know that natural resources are indeed finite, and products should be designed to create the maximum satisfaction or utility with a minimum of material inputs. Although current levels of science and technology do not result in one unit of output per unit of input, it is possible to make much progress by integrating this thinking into product design.

Designers in each industrial sector, for example steel, automobile, chemicals, petrochemicals, pharmaceuticals, paper, machinery, electrical equipment and semiconductors, must assess the raw materials used in making products. Thorough research is then needed to determine what kind of technological revolution is necessary in order to reach equality between inputs and outputs.

2.2.2 The Industrial Clusters Revolution

Second is a revolution in 'industrial clusters' that allows the re-use of waste as raw material. The waste discharged by industry A is used by industry B as an input, the waste from industry B is then used by industry C, and industry C's waste is used as an input by industry D, etc. If industries could be clustered in this way, they could approach Zero Emissions. Attaining true Zero Emissions for all kinds of waste is impossible with today's levels of science and technology.

One example is the use of coal-fired thermal power plants, which discharge large quantities of sulfur oxides (SOx) and nitrogen oxides (NOx) as exhaust, both causes of acid rain. In addition, residues of coal combustion are also produced in the form of ash. Would it not be possible to use this waste as raw material for other processes? Ebara Corporation, an environmental plant manufacturer, investigated this possibility and found that it was able to create ammonium sulfate and ammonium nitrate by irradiating SOx and NOx with an electron beam and adding ammonia. In addition, it is possible to use coal ashes as an ingredient for the production of cement. In short, it is possible to create a cluster of industries that produces much lower emissions by linking the material outputs and inputs of thermal power plants, chemical fertilizer factories and cement factories. Using this theory the company completed an experimental chemical fertilizer plant in September 1997 in the capital of China's Szechwan Province. A plant that uses SOx and NOx as inputs was constructed next to a coal thermal power plant, and today is operating smoothly. Following a request from the capital, the company is also preparing to build a cement factory that will use coal ashes.

To identify useful industrial clusters, input-output analysis using a "waste-resource table" is necessary. In addition, it is important to involve local governments in preparing maps indicating locations and flows, linking companies in the area that discharge waste to those who use it as raw material. It is possible to use such technology as the Internet to expand the scale of such a project to the national level. New approaches such as Life Cycle Assessment and Inverse Manufacturing (Inverse Factories, Inverse Production Processes) are powerful tools to help build industrial clusters. In the field of environmental ISO standards, a part of the ISO 14000 series - environmental management systems and environmental auditing systems - is already coming into effect. The final objective of the ISO 14000 series lies in the life cycle assessment. When this international environmental standard comes into effect, companies will have to bear continued responsibility for their products even after they become waste. Naturally, the arrival of inverse manufacturing designed to turn waste into resources will play an important role here. To get inverse manufacturing under way many factors must be considered, from product design through to disposal. For example, important factors include creating designs that ensure easy dismantling, decreasing of the number of parts used in equipment, choosing recyclable materials, and aiming for durability when developing products.

2.2.3 Taxing the Bad and Cutting Taxes for the Good

The third revolution is fiscal reform on the basis of taxes for the environmental 'bads' and tax reductions for 'goods'. Until now, in order to secure income, governments collected taxes on income earned through fair actions (goods) such as work, savings, and business activity. Income tax, tax on interest and corporate tax are all such taxes. But there has been no specific taxation on undesirable actions harmful to health such as waste discharge, pollution, noise, traffic jams and destruction of the natural environment. It will be necessary in the future to fully revamp existing taxation systems, to reduce taxes on goods such as income and corporate profits (i.e. tax reduction) and on the other hand to collect taxes, and to raise taxation levels (i.e. tax increase) on the 'bads'. Many north European countries such as the Netherlands, Denmark and Sweden introduced carbon taxation at the beginning of the 1990s. This taxation aims at controlling carbon dioxide emissions by curbing the consumption of fossil fuels like oil and coal, through the collection of a carbon tax. In some of these countries, after the introduction of the carbon tax, a decrease of the income tax was attempted while the carbon tax rate was increased. Such taxes on 'goods and bads' are called the "Green Tax System" in Europe. It is possible to decrease waste through the introduction of a Green Tax system.

2.2.4 The Energy Revolution

Fourth is an energy revolution. In order to limit fossil fuel consumption, fuel efficiency must be raised and clean energy must be developed. However, substitute energy sources that can immediately replace fossil fuels are not yet available. A realistic goal would be to first seek technologies and systems that raise fuel efficiency. Technologies such as high temperature air combustion that burns blue flames and allows a 30 percent decrease of CO_2 emissions compared to usual combustion, and the new hybrid cars, are such examples. In the case of the Prius car developed by Toyota, the distance covered per liter of fuel has been doubled (to 28 km) by the combination of a gasoline engine and an electric motor. Innovative technical designs that allow for better fuel efficiency are likely to become more numerous in the near future.

For the development of clean energy sources, the practical use of solar and wind power generation is now possible. If the cost of such energy decreases with increased production brought about by greater demand, the use of clean energy sources could grow more rapidly. The possibility of further technological innovations brought about by growing competition is also an important factor.

The fuel cell also bears great promise. It uses the energy produced by the chemical reaction between hydrogen and oxygen, and only discharges harmless water. Besides Daimler Benz, which has made public its plans to produce and sell a fuel cell car by the year 2004, another German company, BMW, has announced that by the end of the year 2000, it will complete the development phase of its fuel cell car and begin mass production. In Japan, carmakers lead by Toyota, and followed by Nissan, Honda, and Mitsubishi, have all begun the development of cars that will run on fuel cells.

In addition to natural gas (a fossil fuel), hydrogen used in fuel cells could become an important energy source. Hydrogen could provide an escape from fossil fuels because it can be made by improving methanol produced by biomass. If fuel cell use grows to become a decentralized energy source for household electricity in the future, fuel cells may substitute nuclear power, for which the problem of finding acceptable sites is increasingly difficult.

2.2.5 The Lifestyle Revolution

Fifth is a lifestyle revolution. People in developed countries have been pursuing convenient and comfortable lifestyles: in Japan new baths keep bath water hot twenty-four hours a day, household appliances come with standby functions to start instantly, cars and refrigerators are larger than needs require, air conditioners blow air too cold in the summer and heaters are kept too hot in the winter, fruit and vegetables are grown and supplied in any season, engine idling is common, and consumers are flooded with disposable products. Such conveniences, and the pursuit of comfort, waste huge amounts of fossil fuels and resources. However, these lifestyles have brought us to face with environmental and resource limits, and may lead to ecological disaster. We must now change lifestyles to reduce wasteful resource and energy consumption and give priority to more effective use of resources. Especially in large cities such as Tokyo, comprehensive reviews are necessary for transportation systems which promote the use of cars, including implementing restrictions on automobile use in city centers and expanding train (and subway) networks, sidewalks and bicycle paths.

In addition, the tendency for consumers to seek the most luxurious goods is an important factor. Japanese consumers seek luxury goods, and when cars, household electric appliances and computers have succeeded in Japan, they inevitably become best sellers on the world market. While Japan's demanding consumer market has caused a certain amount of pride in Japan, some people are beginning to realize that luxury consumption is damaging the environment. Nevertheless, whereas ten years ago toilet paper that used hundred percent recycled fibers was being used, today toilet paper made from hundred percent virgin pulp has taken a seventy percent market share. This means that forests are being logged just to make toilet paper. As another example, consumers avoid recycled plastic PET beverage containers as they are often slightly scarred when recycled, although the marks have no adverse health affects.

Importing of products causes environmental impacts overseas. In some cases, for one imported log of good quality, ten trees have been cut. Similarly waste occurs on prawn farms when cultivated prawns not big enough for Japanese consumers' tastes are thrown away. Consumers should be aware that the environment is being destroyed overseas for the sake of Japan's demand for high quality goods, however this problem is often unknown to, or even ignored by consumers here. The emergence of a society, which produces Zero Emissions, will be impossible without changes in lifestyle.

2.2.6 Raising the Productivity of Resources is an Issue

From the perspective of corporate management, the key to the Zero Emissions approach lies in knowing to what extent energy and resource productivity can be raised. Basic strategies of corporate management from the industrial revolution until now have gone through three stages: increased labor productivity, improved capital productivity, and finally improved energy and resource productivity. During the twenty-first century as the limits of resources and the environment are reached, the third stage will become important.

During the industrial revolution, when machines replaced human labor that had until that time been the main engine of economic activity, a technological revolution stimulated rapid improvements in labor productivity, which rose further with mass production systems, and still more with successive technological innovations. Production management systems such as the total quality control (TQC) approach introduced in Japan and Toyota's kanban approach (maintaining zero stocks) were enthusiastically introduced in the United States by automobile makers, but again the objective was an improvement in labor productivity.

During the 1990s globalization accelerated and capital efficiency was seriously addressed in Japanese corporate management, with more attention to return on equity (ROE) and to cash flows. This was due to the awareness by management of the extent to which capital investment efficiency could be raised. This was the second stage in Japan.

With the arrival of the twenty-first century, raising the productivity of energy and natural resources is likely to become an important strategy of corporate

Table 1-2. Difference between flow-based and stock-centered economies.

Flow-based	Stock-centered
● Rapid growth	● Stable growth
● Mass production	● Appropriate production
● Disposable products	● Long-life products
● Manufacturing industry	● Manufacture towards service industry
● Export oriented	● Domestic-demand oriented
● Environmental destruction	● Environmental conservation
● Exhaustion of underground resource	● Utilization of above-ground resources
● Centralization of power	● Decentralization of power

management. In the process, new business opportunities will be born, and from those opportunities new business ventures will appear. However, for corporate management, labor and capital productivity cannot be neglected on the simple grounds that corporate strategy has entered the third stage. Labor and capital productivity improvements continue to be necessary in the third stage. However, it will not be permissible to raise labor productivity using methods, which indiscriminately waste energy and resources.

3. TRANSITION TO AN ECONOMY BASED ON STOCKS

3.1 The Limits of Flow-Based Economies

By following the Zero Emissions approach, is it possible for Japan to transform its economy and society to be based on sound material cycles that place only a light burden on environment?

Table 1-2 compares the differences between economies that emphasize flows and those that emphasize to stocks. According to economic theory, flows are sums of added value produced within a fixed period of time. The total value-added produced within a year roughly corresponds to GDP and GNP. On the other hand, stocks are the total of a country's economic assets at a given time. Taking housing for example, the number of new housing starts in Japan each year averages between 1.3 or 1.5 million units (flow), while the present total number of houses in the Japanese archipelago is around 46 million (stock).

The same can be said with reference to cars. Annual production (flow) has recently been around 10 million cars per year. On the other hand the number of cars owned (stock) is around 73 million, about seven times this number. Flows include the assets that have not been consumed during a given year, and are added to the stock of the next year.

The Japanese post-war economy was a typical economy based on flows. This kind of economy seeks to maximize value-added and is supported by mass-production and mass-consumption. An economy, which has not reached the mature stage, lacks social capital such as housing, roads, railroads and bridges, and equipment capital such as factories and office buildings. Because the cumulative wealth (stock) of society is in short supply it is effective to prioritize flows in the economy.

In post-war Japan, many factories and houses stood in ruins due to the damage suffered during the war. With the total destruction of social capital such as roads and railroads, coupled with heavy damage of other capital, the economy had fallen into an extreme stock shortage. Strong growth was desirable to make up for this stock shortage, because with a high growth rate the stocks would accumulate. Post-war Japan was rebuilt though this model of development.

Table 1-2 illustrates flow-based economies characterized by high growth, mass production, disposable products, manufacturing industries, export leadership, environmental destruction, resource depletion, and centralization.

Mass production is necessary in order to maximize value-added. Production of disposable goods is a powerful method to reach this goal. It is possible to increase value-added in the economy by if a cycle can be created in which goods that are thrown away after use, requiring more production, mores sales, disposal, and more production. While goods such as cars and electric household appliances are not exactly disposable, value-added in the economy can be maximized with frequent model changes that have the effect of making perfectly usable goods look old fashioned. Moreover, it is possible to increase the value-added by production in excess of domestic demand and exporting the surplus.

In other words, an economy based on production to meet the original demand for goods is limited, and in order to maximize the value-added to its highest potential, wasteful production and consumption become necessary. This is why during the period of high economic growth in Japan advertising copy praised the virtues of consumption and encouraged extravagance.

But Japan's flow-based economy has produced mountains of waste and led to mountains of negative impacts, such as waste disposal problems of crisis proportions, accelerating environmental destruction and resource depletion, and the detection in the environment of man-made chemicals hazardous to health such as endocrine disrupters. History has shown that flow-based economies have not been sustainable because they are limited by environmental factors and resource supply.

3.2 Progress based on Appropriate Production

Key concepts in the stock-centered economies include stable growth, appropriate production, durable products, conversion from manufacturing to service industries, domestic demand, environmental preservation, utilization of above-ground resources and decentralization (Table 1-2). Every one of these words contrasts with the key concepts of flow-centered economies.

Japan could be described as having a mature economy today, because it has accumulated a substantial level of stocks. For example, although Japan may compare unfavorably with the United States or Europe on housing quality, its housing stock today exceeds one house per family. Nationwide sufficient social capital exists in the highways, high speed trains and bridges, and in facilities such as cultural and concert halls, museums and gymnasiums. In fact there are now calls for a halt to new construction. Every household now boasts many appliances and equipment such as televisions, refrigerators, air conditioners and personal computers, and wardrobes are filled with more clothes than one could ever possibly use. In rural Gunma Prefecture it is not unusual to find farm families possessing five or six cars. People that need only one car often get more than they need.

In the future Japan must make its existing stock last longer and use it efficiently. For example, housing improvements should come from improving quality and durability rather than demolition and rebuilding. This path will lead to a better quality-of-life, to the preservation of the environment and to efficient use of resources.

A characteristic of stock-centered economies is that growth is not sought just for the sake of growth. Of course, the necessities of life must be produced, but there is no need to render as obsolete the products that can still be used, and no need to promote mass-production to the extreme of encouraging waste. Rather than mass production, it is appropriate production that is needed now. Key drivers of appropriate production are demand for replacement or rebuilding. New products replace old ones only when they need replacing. With housing, the demand for rebuilding replaces the demand for new housing starts.

Japan has already reached an ownership rate of one house per family, but the population will begin to decrease from the year 2010. Naturally, the demand for new homes will decrease and the demand for housing renewal such as rebuilding will grow. In consideration of this, measures and guidance are needed to encourage a transition towards good quality and comfortable homes. With this objective in mind, it is necessary to lay out the conditions that will promote the formation of a good quality housing stock, for example through systematic preferential taxes, like the ones put into effect in post-war America, which allowed a total tax deduction of housing loan interest.

For durable consumer goods like cars too, considering that every household already owns a car, production in the future should be only at the rate of the replacement demand. However, if gasoline-powered cars are to be replaced with low emission or zero emission vehicles, for environmental reasons it may become necessary to break this rule. Of course, it is necessary to have some flexibility in responses and policies.

3.3 Dramatic Changes in the Approach to Product Development

Corporate thinking about product development will naturally have to change. In flow-based economies, the disposable products boost corporate profits. On the other hand, in stock-based economies the development of durable products should be profitable. The current practice in Japan of rebuilding wooden houses after approximately 25 years is a common example of a flow-based economy. In contrast, in a stock-centered economy, houses will be built using the same wood that can be habitable for a century.

Housing makers here have been rebuilding houses at the rate of four times a century, but this frequency should be reduced to just once. While some companies will resist this suggestion, housing makers of the new stock-centered economy will see things differently. It will be possible to build durable houses that can last a century, and to maintain profits by increasing income from services such as repair and alterations as the role of these companies changes to one of 'family doctor' for houses. Similar approaches can also be applied to industries making durable consumer goods such as cars.

The process described here is a transition from manufacturing to service focus for industry. Manufacturing industries will continue to make products, but management thinking that business is only about producing and selling product will no longer be acceptable. Corporations will have to take responsibility for their products "from the cradle to the grave."

Income can be earned from the production and sale of long-life goods, and also from the service industry for the repairs and replacement of parts during the life of a product. When the product life has ended and a product is disposed, a closed system will become necessary in which the manufacturer is responsible to accept this product, dismantle it to remove components that can be re-used, and recycle materials which are then used in the production of new products. While helping improve resource productivity, closed system operations will help to de-materialize the manufacturing industry.

3.4 The Use of Above-Ground Resources by Developed Countries

Next it is necessary to consider above-ground resources -- an important concept for stock-centered economies. This paper has referred to the issue of resource depletion, but to where do these resources disappear? In developed countries such as Japan, above-ground resources have accumulated in the form of roads, railroads, bridges, high rise buildings and individual houses. If methods to allow the extraction of these resources once they have been accumulated in other forms are found, it will be possible to recycle and use them for very long periods of time. In addition, the re-use of above-ground resources saves enormous amounts of energy. For example, only one-third the energy is required to produce one tonne of crude steel by melting scrap iron in an electric furnace, compared to melting iron ore in a blast furnace. For the recycling of scrap copper this ratio drops to one tenth. In the case of aluminum, sometimes called 'canned electricity,' the recycling aluminum scrap requires only 1/190th the electrical power consumption required to refine raw material, representing an enormous energy saving. Similar energy savings can be found in the case of other metals.

Japan's annual iron production from iron ore is around 100 million tonnes, but ten times this amount has accumulated in stocks as above-ground resources. The United States and the European countries are similarly rich in above-ground resources. In these countries it will be necessary to extract these above-ground resources efficiently, and to conserve the dwindling underground resources for developing countries that still need them, and also for the generations to come.

Also, in order to use the above-ground resources efficiently, it is necessary to prepare detailed above-ground resource inventories, stating, for example, how much of each resource has been used in man-made structures, such as a high rise buildings and how many years the structure is expected to last. Plans should be developed to use all the above-ground resources stored in these structures once they outlive their use.

4. CREATING AN ENVIRONMENTALLY SOUND MARKET ECONOMY

4.1 Laws and Systems to Support a Society with Sound Material Cycles

This paper proposes that in order to build the recycling economies based on the Zero Emissions approach, the flow-centered Japanese economic development pattern must be changed to a stock-centered one. Is there now in Japan momentum promoting such change?

Concerning this question it is important to note that in the context of international landmarks events such as the Earth Summit in 1992 and the Kyoto Conference on climate change in December 1997, at a rapid pace the government of Japan has been establishing the legislative infrastructure for a society built on sound material cycles.

The Basic Environment Law, which could be considered the 'constitution' for the country's environmental policy, was established in November 1993. Because this law only expresses in an abstract way the philosophy of environmental preservation, the Basic Environment Plan based on this law, aiming at the general promotion of governmental policies, was adopted by the Cabinet in December 1994. During the June 1997 ordinary diet session, the Environmental Assessment Law was enacted in an effort to expand upon the spirit of the Basic Environment Law. Two more laws, namely the New Energy Law and the Law Promoting Measures Against Global Warming, both aimed at the realization of Japan's commitments under the Kyoto Protocol (6 percent reduction of greenhouse gas emissions by about the year 2010), were adopted in 1998.

Complementing the above during the 1990s, laws to promote recycling were rapidly introduced, for three main reasons: resource depletion, bringing with it the necessity to use resources more carefully and to recycle waste; the need to convert harmful waste into useful resources, in order to prevent pollution resulting from waste disposal; and the severe shortage of waste treatment plants that has arisen out of the mass-production, mass-consumption, mass-disposal 'one-way' economic system. In particular, as waste is growing rapidly this third reason is an even more urgent problem today.

In short, the 'one-way' approach of the past that called for new waste disposal sites to be built whenever there was waste, has come up against a huge barrier. A hundred-eighty degree turn in thinking is needed, as we are entering the era of a society that strives to produce no waste, and when waste cannot be avoided it must be recycled. We have come to the point where no other choice is permissible.

The Public Cleansing Law relating to waste disposal was enacted in 1954 and implemented for pollution treatment in cities. Later, as a result of economic development, the amount of waste discharged by companies grew rapidly, leading to the need for the Waste Disposal and Public Cleansing Law which was introduced in 1970 and enforced starting in 1971. That law was based on the 'one-way' thinking of waste transport from source to the disposal facility for treatment. By the 1990s as environmental problems began to be taken more seriously, and recycling attracted greater public attention, the law was substantially revised in 1991 and 1997 to include ideas about the building of recycling societies. In particular, in the 1997 revision improvements included requiring efforts by waste emitters to recycle and limit waste generation; making mandatory public disclosure and consultation with local communities when waste treatment facilities are to be built; and broadening a 'manifest' reporting system to include all kinds of waste in order to prevent illegal dumping.

During the 1990s Japan's first law focused specifically on recycling was established -- the Recycling Law (Law for Promotion of Utilization of Recyclable Resources) of 1991. This law aimed to increase the recycling rate at the raw materials stage, promote the supply of easy to recycle products and recycle by-products discharged by factories.

The Containers and Packaging Recycling Law was introduced in 1995. The recycling of aluminum, steel and glass containers, paper beverage containers and plastic PET bottles was put into practice in April 1997 on the basis of this law. It will be put completely into practice for all paper containers and packaging other than of beverages, and for all the plastic containers and packaging other than PET bottles in April 2000. The most important characteristic of this law is that it divided the obligations three-ways, involving municipal governments (to collect waste separately), consumers (to separate waste) and producers (to recycle products).

The Electrical Appliances Recycling Law (Law on Specified Electrical Household Appliances Recycling) was enacted and put into effect in 1998. The recycling of the four categories of targeted appliances -- televisions, refrigerators, washing machines and air conditioners -- will be put into effect starting in the year 2001. In the future other appliances may be targeted.

A bill to promote the recycling of construction site waste, which is being considered by the Ministry of Construction, also deserves mention. With the administrative reforms to occur in 2001, the authority for waste treatment will be transferred from the Ministry of Health and Welfare to the Environment Agency, which will be promoted to the status of a ministry. As Japan conducts an overall review of the existing Waste Treatment Law, it is working quickly to set up the legislative infrastructure to create a recycling society, as was done in Germany with its Closed Substance Cycle Waste Management Act.

Such initiatives provide excellent conditions for the building of society that has sound material cycles and is based on the Zero Emissions approach.

4.2 The Limitations of a Laissez-Faire Market Economy

What kind of economic system may be desirable to build a stock-centered recycling society? During the twentieth century, two economic systems existed in opposition: capitalism which was supported by the market economy, and socialism which based on planned economies such as in the former USSR. Which model is optimal for a society with sound material cycles and for environmental preservation?

The market economy has been the most suitable system for optimal resource distribution. So how is it that market economies have depleted resources and caused degradation of the environment? The most important reason is that environmental costs have not been properly accounted for. The reason for this is that the regenerative power of nature was considered relatively large compared to the market size, so until now there was no perceived threat to human living conditions comparable to today's environmental destruction. Clearly, a pure market economy which is close to laissez-faire in approach is not a desirable system because it wastes resources and promotes environmental degradation.

What is the situation in the case of planned economies? They have presented an even greater threat of environmental degradation and waste of resources than market economies. During the late 1980s and early 1990s, as the socialist systems of the former USSR and eastern European countries collapsed, Western countries were shocked as the reality of the environmental degradation in those countries came to light. Due to the use of thermal power plants fueled mainly by coal, air quality around industrial areas of countries such as Czechoslovakia, Poland and Hungary, suffered from severe air pollution. Pollution from chemical waste discharged into rivers and lakes was devastating. The destruction of forests caused by acid rain also drew great attention. Further to this, nuclear pollution has become a serious problem in the former USSR, where some sixty locations are severely polluted by radioactivity. The main cause of this pollution was poor management of radioactive isotopes used for industrial purposes. Planned economies are systems where resource distribution is not entrusted to the market, but to bureaucrats under authoritarian governments, and experience has shown that these do not function well.

4.3 Image of an Environmentally Sound Market Economy

This discussion leads us to a realistic model whereby an environmentally sound market economy can be constructed and supplemented with adjustments to the laissez-faire market economy that introduce environmental costs to the market.

With this as an objective, what are the roles of corporations, consumers and the government?

First, for corporations the primary requirement is the introduction of environmental costs into management decision-making. The old management doctrine that sees environmental costs as external costs as opposed to personnel and other expenses that are internal, is still prevalent in Japanese corporations. But the costs involved in making products, the investment costs that corporations must bear for equipment to reduce the environmental burden at the point of production, the costs of disposal and recycling of waste from production processes, and the costs of research and development of environmentally friendly products should logically be counted as internal costs and be counted in the cost of production.

The problem therefore is how to deal with corporations that consider environmental costs as external and refuse to count them as internal costs. If the situation is left unchanged, corporations that are not counting environmental costs in their production may offer cheaper products and benefit from an unfair advantage over corporations that do bear responsibility for the environmental costs of production. Legislative or institutional measures are needed to keep those corporations from the market. One possible method is an aggressive approach involving penalties such as with an environmental dumping tax. For this purpose, each economic sector must estimate environmental costs per unit of product.

Although the Fair Trade Commission in Japan has been regulating activities such as unfair discounting, regulation has not yet addressed the issue of environmental dumping. Today new rules should be instituted to prevent market entry by corporations that do not count environmental costs. Internalization of environmental costs must become the corporate norm in the twenty-first century.

4.4 The Main Role for Green Consumers

Second, consumers must change. Twentieth century consumers gave importance to economic rationality (homo economicus). They typically chose without hesitation the cheapest between two products of the same quality.

On the other hand, consumers who support an environmentally sound market economy give importance to the environment (green consumers). They are a new breed of consumers who will tend to purchase a product if it is environmentally friendly, even if it is more expensive. Generally, consumers in countries where the economy is in the developmental stage show an overwhelming tendency to make purchase decisions based on price rather than on environmental standards. As the economy reaches a more mature stage, consumers will start to feel slightly undecided. While thinking that they probably ought to buy the environmentally friendly products even if the price is slightly higher, when actually making a purchase the tendency to choose the cheapest, easiest to use products is still a deep-rooted behavior. However, consumers in countries with mature economies are more likely to escape from such hesitation and give importance to the environment.

Figure 1-4. Interest in global environmental issues (Source: the Prime Minister's Office, November 1998).

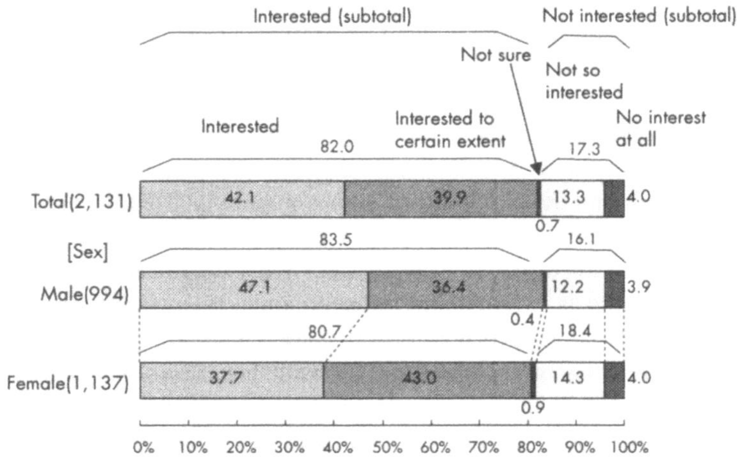

Figure 1-5. Items considered in everyday life.

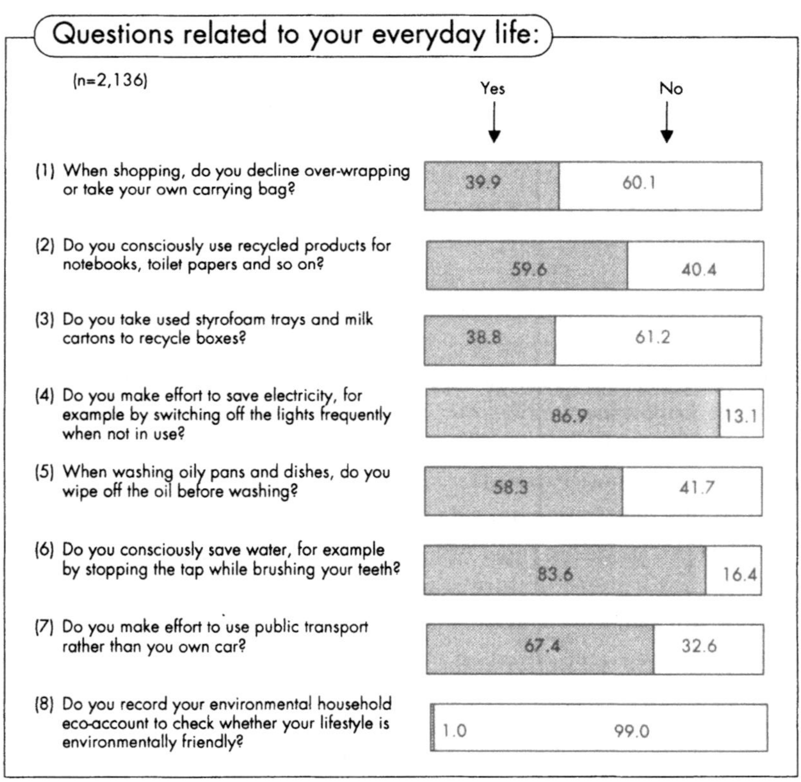

Figure 1-6. What consumers are concerned, from a point of view of enterprises.

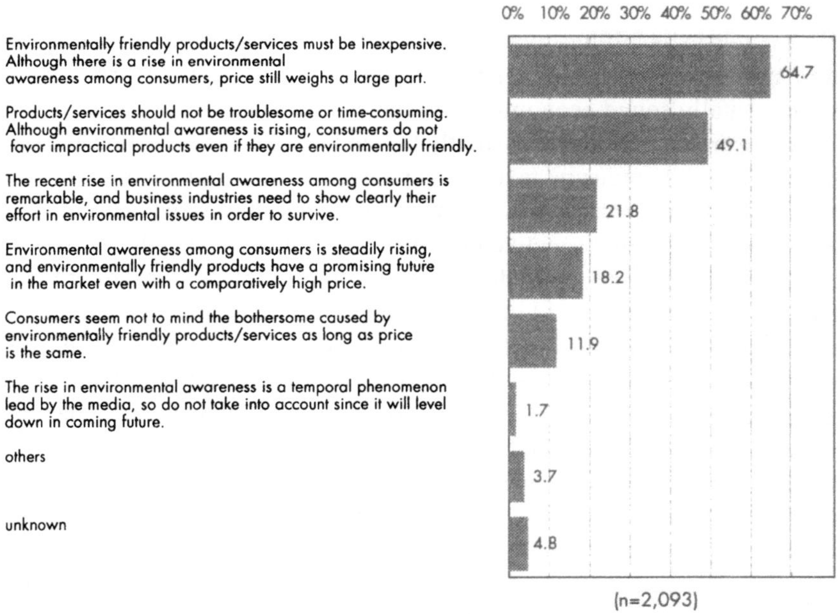

For example, in some European countries such as Germany, the Netherlands, the U.K. and Sweden, a steadily increasing number of consumers are joining a system in which they pay higher than average electricity charges, with the difference in price being allocated to the development of energies such as solar and wind power. Some consumers insist on more expensive clean energy, rather than the cheaper fossil fuel energy. Here we can see appearing the image of the new twenty-first century green consumers.

Today Japan's consumers probably stand at the end of the second stage mentioned above. According to an environmental survey carried out by the Prime Minister's Office and the Tokyo metropolitan government, many consumers express concern for the environment (Figures 1-4 and 1-5). However, their efforts are still slow when it comes to making purchasing decisions, refusing excessive packaging, carrying their own shopping bags, or returning disposable polystyrene trays to the supermarket for recycling. Table 1-6 features the consumer's product buying attitudes as seen by producers. It shows that while environmental concern does exist, consumers still purchase cheap and easy-to-use products.

Nevertheless, there is certainly a growing tendency among Japanese consumers to choose environmentally friendly products. As environmental awareness grows in the future, it is expected that Japanese consumers too will rapidly become green consumers.

4.5 Promoting Regulatory Incentives Through Tax Reform, etc.

Third, the role to be fulfilled by the government in order to build an environmentally sound market economy is also important. Tax system reform leading to environmentally sound economic activities is important. As mentioned, new tax systems are urgently needed that strictly tax environmental destruction and resource consumption while reducing taxes on personal and corporate income. In addition, in the twenty-first century it will be important to introduce fixed taxes on all virgin resources, together with incentive measures that will encourage the use of recycled materials, to address resource depletion and to promote above-ground resource recycling. These changes will make recycling more profitable and attractive.

Other possible measures include establishing stricter control over the levels of chemical discharge hazardous to health, over the levels of noise pollution and rush hour traffic. On the other hand, incentives such as subsidies and special depreciation rates are also necessary to support the new technologies for environment preservation and fuel efficiency improvement. In these ways corporations can be allowed to compete freely in a market economy within a framework based on legislation and economic measures that promote environmental preservation.

An environmentally sound market economy will probably emerge through the new roles of consumers, corporations and the government. During the transition to this new economy, strong resistance can be expected from every vested interest group that profits from the existing systems and sees any changes as threats. They will maneuver to maintain the present situation by pressuring politicians and winning over bureaucrats.

But if the systems which may have been efficient in the past cannot meet the needs of today, the new winds of change will inevitably blow. The winds are steadily gathering speed for the new era of an environmentally sound economy based on sound material cycles.

Chapter 2

ENERGY AND NEW DEVELOPMENT PATTERNS

HARUKI TSUCHIYA
Research Institute for Systems Technology

1. INTRODUCTION

During the twentieth century, humanity expanded its activities through the application of science and technology. The results are visible through the rapid increase of the global population and the achievement of higher standards of living in industrialized countries. As lifestyles based on mass production and mass consumption become the norm in these countries, the mass consumption of resources and energy has also become chronic. The effects of these trends have not been limited only to those countries; they have also had significant effects upon developing countries. The global population has come to rely on energy in abundance. The need for a large energy supply in turn has made us depend heavily on fossil fuels, which emit carbon dioxide. The risk of global warming caused by the buildup of carbon dioxide in the atmosphere has therefore become a hot issue of debate.

The IPCC (Intergovernmental Panel on Climate Change) which was organized by the United Nations and World Meteorological Organization compiled their Second Assessment Report at the end of 1995. This report disclosed scientific data which warned that global warming caused by human activities might have already begun.

In December 1997 the Third Session of the Conference of the Parties (COP3) to the UN Framework Convention on Climate Change was held in Kyoto. The outcome was the Kyoto Protocol in which developed countries committed themselves to decrease their global warming gas emissions by an average of 5.2 percent from 1990 levels in the period 2008 to 2012. This is a legally binding protocol that is to be signed by the participating countries. It has been likened to a first "disarmament" treaty of humanity's global activities of production and consumption.

K. Matsushita (ed.), Environment in the 21st Century and New Development Patterns, 47–67.

The protocol also gave justification to the ideas expressed in Limits to Growth, the report submitted to the Club of Rome by D. Meadows in 1972. The ideas were not fiction; they had proven themselves true. With the Kyoto Protocol, humanity has at last begun to step forward to limit its harmful activities.

As the twentieth century has been called "the century of industry," the twenty-first century should perhaps be called "the century of environment." Under these circumstances, the concept of New Development Patterns should include an emphasis on the importance of decreasing our heavy dependence on fossil fuels, improving energy efficiency, and of increasing the supply of decentralized and renewable energy. This report summarizes the main points of such discussions, including the future of global energy consumption, improvements in energy efficiency, renewable energy, decentralized energy supply, the relationship between decentralized energy systems and economics, and New Development Patterns.

2. GLOBAL ENERGY FUTURE

It has been predicted that by the middle of the twenty-first century, the global population will reach 9 to 10 billion people. This means the current global population will almost double in the next fifty years. If standards of living remain at their present levels, clearly energy consumption will also double by the middle of twenty-first century. If the global economy grows considerably, standards of living will become substantially higher. If levels of energy efficiency remain constant, the global energy consumption may quadruple, given predictions of the growth of GDP per capita combined with population growth. However if energy efficiency improve to twice the current levels, the future energy demand could be controlled at twice the level of today.

Below is one scenario for the middle of twenty-first century.

(a) The global population increases twofold.

(b) The improvement of energy efficiency will contribute to a fifty percent decrease in energy consumption per capita compared to today's level in developed countries. (The current rate is 5 to 10 kilowatts, and is assumed to improve to 2.5 to 5 kilowatts in the period 2050-2100.)

(c) The rate of energy consumption per capita in developing countries will be from 2 to 10 times higher than today's level. (Currently 0.2 to 1.0 kilowatts, it is assumed to be 1 to 2 kilowatts per person in the period 2050-2100.) One kilowatt per capita is considered the level of energy required to maintain a minimum standard of living.

(d) The global energy demand will be twice that of today's level.

It is difficult to ascertain whether or not this simple scenario is realistic, as many variables are involved. If the above prediction is correct, and should we continue to depend heavily on fossil fuel, global warming is inevitable. It is important for society to begin to use the energy sources that do not emit carbon dioxide-i.e., renewable energy sources. Let us now consider the possibilities for increasing the supply of renewable energy resources in the next fifty years.

2.1 Energy Scenarios in IPCC Second Assessment Report (1995)

The energy scenarios in IPCC Second Assessment Report are shown in Table 2-1. The population in the year 2100 is estimated to range from 6.3 to 17.6 billion and the economic growth rate is assumed to be 2 to 3 percent during the period. The energy supply is anticipated to switch towards solar energy from oil and coal.

Table 2-1. Six Scenarios by the IPCC.

Scenario	Population in 2100 (billion)	Economic Growth Rate (%) 1990-2025 1990-2100	Energy Supply
IS 92 a, b	World Bank (1991) 11.3	2.9 2.3	Oil 12,000EJ, Gas 13,000EJ Solar 7.5cents/kWh Biomass $70/barrel 191EJ
IS 92 c	UN Medium Low 6.4	2.0 1.2	Oil 8,000EJ, Gas7,300EJ Nuclear Cost decreases 0.4%/year
IS 92 d	UN Medium Low 6.4	2.7 2.0	Oil 8,000EJ, Gas7,300EJ Solar 6.5cents/kWh Biomass $50/barrel 272EJ
IS 92 e	World Bank (1991) 11.3	3.5 3.0	Oil 18,400EJ Gas same as a,b Nuclear phase out in 2075
IS 92 f	UN Medium High 17.6	2.9 2.3	Oil and Gas same as e. Solar 8.3 cents/kWh Nuclear cost increases to 9 cents/kWh

2.2 LESS Energy Scenario (IPCC Second Assessment Report)

The IPCC Second Assessment Report includes the Low CO_2 Emitting Energy Supply System (LESS Scenario) as part of the energy system mitigation option. This is a thought experiment to show the possibility of decreasing CO_2 emissions. This scenario combines energy efficiency improvements, renewable energy, the use of low CO_2 fuels, and the use of nuclear power. It emphasizes aggressive use of natural energy, especially photovoltaics (PV) and biomass.

The assumptions are as follows.

* Population: 9.5 billion (2050), 10.5 billion (2100)

* GDP: in 2050, 7 times 1990 levels (5 times in developed countries, 1.4 times in developing countries); in 2100, 25 times 1990 levels (13 times in developed countries, 70 times in developing countries)

There are two scenarios. First is the Low Energy Demand Scenario in which primary energy demand increases 1.9 times from 1990 to 2100. Net demand will remain as in developed countries, but will be 4.4 times in developing countries. Second is the High Energy Demand Scenario. Primary energy demand increases to 3.7 times from 1990 to 2100. It remains the same as in scenario IS92 a.

The various cases for calculations are as follows (Figure 2-1):

BI: Biomass intensive NGI: Natural gas intensive CI: Coal intensive

NI: Nuclear Intensive HD: High Energy Demand

Figure 2-1. The LESS scenarios.

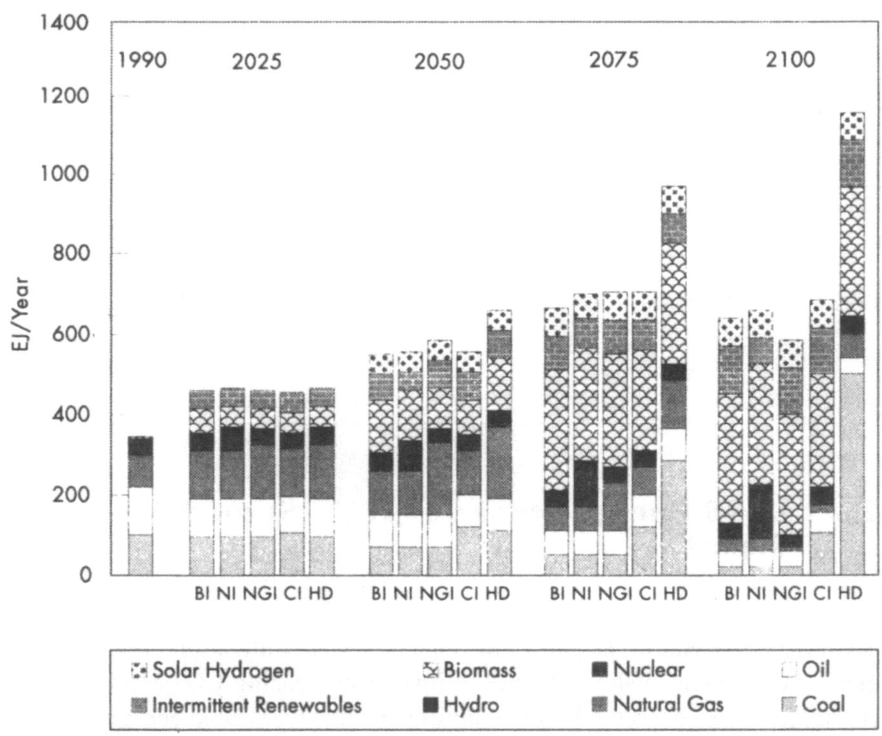

Table 2-2. Energy Supply Scenarios in 2050.

Primary Energy Supply	1990 data	IS92 a Scenario	WECC/ IIASA B	WECC/ IIASA EcoDriven	Shell Sustain Growth	Shell Dematerialization	IPCC LRSS BI
Coal(EJ)	96.9	356	179	66	184	150	56
Oil(EJ)	142.9	153	179	121	148	121	75
Gas(EJ)	76.0	143	210	182	140	182	120
Fossil fuel Total(EJ)	315.8	652	568	369	472	453	251
Nuclear(EJ)	22.8	87	121	22	94	62	11
Renewable(EJ)	31.9 (+41.6)	252	141 (+56)	195 (+55)	619 (+61)	207 (+61)	312
Total (EJ)	370.5 (+41.6)	991	830 (+56)	586 (+55)	1185 (+61)	722 (+61)	574
Per Capita Energy Demand(kW)	2.09	3.31	2.77	1.95	3.95	2.41	1.92
Nuclear+Renewable Ratio (%)	14.76	34.21	31.57	37.03	60.17	37.26	56.27
Renewable Ratio(%)	8.61	25.43	16.99	33.28	52.24	28.67	54.36

(Unit: global EJ/year)

2.3 Comparison of Several Energy Scenarios

The comparison of scenarios with a predicted human population of 9.5 billion in 2050 is shown in Table 2-2. Here the WEC/IIASA scenarios (Scenario B and Ecologically driven) and the Shell Scenarios (Sustainable growth and Dematerialization) are also included.[1] The rates of nuclear energy and renewable energy are calculated in Table 2-2.

This table shows a wide range of estimated energy demand in 2050, with increases from 1.5 times to 3.2 times compared with 1990. The rate of energy consumption per capita in 2050 is between 1.9 and 3.95 kilowatts. The ratio of non-fossil energy supplied (nuclear + renewable) ranges from 31 to 60 percent in 2050, while the ratio of renewable energy consumption is from 17 to 54 percent in 2050. These numbers show how much energy experts expect from renewable sources in 2050.

The share and actual size of the renewable energy supply is projected to increase in any of the given scenarios. It is clear that the energy experts expect increases in the renewable energy supply in order to mitigate global warming. Opinions differ on the assessment of nuclear power. Associated with nuclear power are three main problems: nuclear proliferation, safety, and nuclear waste disposal.

3. IMPROVING ENERGY EFFICIENCY

An ordinary analysis about energy problems begins with an evaluation of energy demand and supply. The logic is as follows: the energy demand grows along with increases in the GDP and improvements of the standard of living. At the same time, exponential increases in the energy demand call for larger supplies of energy, as may be satisfied through the use of coal, oil, natural gas and nuclear power. Given that experts predict the depletion of oil supplies in the next forty years or so, non-oil energy sources must be discovered and used on a global scale.

Many energy experts had similar discussions after the first energy crisis in 1973. They were generally involved in research relating to energy supply industries and their expertise was mainly in the supply of oil and coal. It would be difficult to expect them to deal with new situations while they had such a strong grounding in traditional thinking about energy.

However different ideas appeared in the late 1970s and 1980s. These included: Soft Energy Paths (1977) by Amory Lovins (now at the Rocky Mountain Institute) and the End-Use Oriented Approach by a group in the Center for Energy and Environment at Princeton University (1982). Their approach pointed out that people do not need energy itself; rather, they need the services provided by energy. In contrast, the conventional approach was to feed the ever-growing energy demand by exploiting large energy sources. This approach was called the "supply oriented approach." These two different approaches are sometimes named as the "top-down" and "bottom-up" models, which were explained in the IPCC Second Assessment Report.

According to the End-Use Oriented Approach, by focusing on the end-uses of energy, one would find much room to improve energy efficiency. If energy were supplied in order to satisfy demand from decentralized renewable energy sources, the results would be practical and highly efficient.

This End-Use Oriented Approach became widely known when the books Energy for Development and Energy for A Sustainable World authored by the Princeton group were published by the World Resources Institute in Washington in 1987. The key messages of these books are that our world's use of energy is inefficient and there is much room to improve efficiency. The books helped destroy the myth that emphasized the need to find other large energy resources because oil would be depleted in the twenty-first century.

Let us review what has happened. Energy prices skyrocketed during the oil crises of 1973 and 1979, fostering energy conservation worldwide. It became clear that the price of energy inevitably drove energy conservation. Amazingly, energy demand in developed countries did not grow while the economy flourished in the first half of 1980s. The energy demand elasticity to GDP, which shows the ratio of energy growth to economic growth rates, was very low and the iron link between energy and economy appeared to melt down. The real potential of efficiency improvement was widely recognized. In Japan the primary energy demand elasticity to GDP was minus 0.1 from 1979 to 1986; energy demand decreased by 0.1 unit while the economy actually grew by 1 unit in this period.

The spread of energy conservation created a new situation, resulting in a decrease in energy demand and the subsequent fall of oil prices. However the lower prices actually caused practices to revert back to the wasteful ways of before the crises. Energy demand began to increase again by the end of the 1980s. The primary energy demand elasticity to GDP was 1.14 between 1986 and 1995 in Japan, when energy demand growth was higher than the GDP growth and the concerns about energy conservation receded. The history of the relationship between energy prices and energy efficiency shows that motivation to conserve energy is correlated to energy prices.

However many energy efficient technologies already exist, and it is not only the price of energy that drives advances in energy efficiency. If the goals of society are clearly articulated, many policies and measures could be used to promote energy conservation. Today many energy-efficient technologies already exist that could be economically viable; the problem is that they are not used on a large enough scale. Some perspectives on energy efficiency are included below.

3.1 Energy efficiency

Dr. Norgard at the Denmark Institute of Technology showed that energy efficiency could be subdivided into device efficiency, social efficiency and life style efficiency.

Energy Efficiency = Device efficiency x Social system efficiency x Life style efficiency.

Device efficiency here is considered to be the purely technical efficiency of an end-use device. It is the ratio of effective energy service compared to the amount of energy input, such as when we use a refrigerator, automobile, heat pump or insulation to fill a need. Innovation and invention have continually upgraded and improved these products, but improved designs are not as widespread as they could be. Low energy prices create no motivation to improve, or users lack the needed information.

Social system efficiency is energy efficiency inherent in the social structures, traditions and customs in society. If a society has only one way of doing something, then people have little choice. In Los Angeles, the poor state of public transportation compels people to use cars.

Life style efficiency is related to the efficiency of an individual's way of life. Some people prefer to wear light clothing indoors, while consuming enormous amounts of energy for their heating houses in winter. On hot summer days some buildings and department stores blast cool air from air conditioners at their entrances.

The combination of the three efficiency factors determines the energy demand in a society. The energy consumed by a multitude end-users adds up to enormous amounts of energy demand. A related example of energy efficiency follows.

3.1.1 End-use Device Efficiency

Energy-consuming devices are at work in factories, in transportation and in households. The energy demand in the industrial sector accounts for nearly half of the primary energy demand in Japan. Most of this energy is consumed in the production of materials in such energy intensive industries as iron and steel, paper and pulp, cement, and the petroleum industry. The energy efficiency is shown as the energy intensity, i.e., the energy input required to produce one unit of material output, for example, one ton of steel or paper. If the energy intensity is low, efficiency is high. The energy intensity is normally the same anywhere because similar processes are used in most factories. International comparisons of energy intensity do show however, differences of a factor of 1.5 or 2 between countries. The gaps between countries show that room for improvement does exist. In many cases industry is very sensitive to energy costs, so if energy prices are low, energy intensity becomes less of a concern. In spite of the fact that highly efficient processes and devices have been available for quite some time, many industries will still not make the effort to improve efficiency because they cannot quickly recover their investment. Substantial investments were made into energy efficiency in 1970s, but such enthusiasm has declined as a result of low energy prices today.

Recently the concept of a society based on sound material cycles has attracted attention, with the recycling of materials seen as contributing to energy efficiency, especially in cases involving materials such as aluminum and other metals.

The automobile provides one example of the potential for efficiency improvement in the transportation system. To carry one unit of cargo or one person, the automobile consumes more energy than public transportation such as trains and buses. The energy efficiency of the car is important because it accounts for the biggest share of energy demand of transportation systems. Automobile manufacturers in the past did not pay much attention to efficiency.

The situation is changing, however gradually. The hybrid car released by Toyota in 1997 runs 28 km per liter of gasoline (65.7 miles per gallon). It is twice as efficient as a conventional 1500 cc sedan in the same class. The term hybrid refers to the combination using a gasoline engine and an electric motor. When the car starts, the gasoline engine remains silent while the energy from the battery drives the wheels using an electric motor. At the speed of about 30 km per hour, the gasoline engine starts to drive the car. Extra energy is converted to electricity and charges the battery. When the car stops moving at intersections, the gasoline engine automatically stops. When the brake is applied, the momentum of the car is sent to the generator and the electricity produced charges the battery. It was previously thought that the hybrid concept was possible only in large vehicles because of the required size and weight of the complex components. However, technology that was thought to be only a dream became reality, and in 1997 the car was on sale for only about U.S.$4,000 higher than a conventional car. By mid 1999 more than 20,000 vehicles had been sold. If this hybrid concept were adopted by Japan's entire automobile market, 10 percent of the country's total CO_2 emissions would effectively be eliminated.

A typical refrigerator consumes nearly 20 percent of the total electricity demand in a household, and runs year round. Thicker insulation and more efficient pumps have been credited with improving efficiency, but the additional costs have made manufacturers hesitate to compete in the market. As a result, in Japan refrigerators have become less energy efficient than those manufactured in the 1980s.

The energy efficiency associated with lighting improves tremendously if one uses fluorescent instead of incandescent lamp lighting. Moreover, highly efficient fluorescent lighting that uses inverter control has become available on the market.

Building and housing insulation decreases heating demand. The most effective way to promote insulation is to improve the building standards, which have been gradually changed to require better insulation.

As described above, the energy devices in today's marketplace are not the best possible in terms of energy efficiency. Many still have much room for improvement through technological development and improvements in industry standards. Many ways to improve energy efficiency already exist, but have not yet been widely applied.

Life Cycle Assessment (LCA) is effective as a comprehensive method of analyzing energy efficiency. LCA shows the energy consumption through production, usage, and disposal, throughout the lifetime of a particular product. Eighty percent of energy consumed by a car through its life is used while it is running. This tells us that fuel economy is most important for a car as a means to improve efficiency. An LCA of a house reveals that heating and cooling accounts for 60 to 80 percent of energy consumed by a house during its lifetime (though this also depends on the local climate), so for houses insulation is the key to efficiency.

3.1.2 Energy Efficiency of Social Systems

Energy efficiency of a society is usually invisible. Public transportation systems such as trains and subways are not only energy efficient; they also emit lower levels of pollutants into the atmosphere than cars. It is often a more punctual way to travel in an urban area than by car. In recent years many local governments promote policies to keep cars out of central areas of the city.

If a society is interested in introducing sound material cycles, paper and metals will be recycled more, and the energy input required for material production will decrease. This is the significance of the energy efficiency of social systems.

There are other aspects of a social system's energy efficiency. Consider an architect designing an air conditioning system for a large building. If he chooses large heating equipment and ducts, the cost of building becomes higher and probably his fee will be larger. The operating cost for the building will also be high. In contrast, if the architect's remuneration were linked to the energy efficiency of the building design, these types of designs would become more popular. More attention is needed to such mechanisms in society.

3.1.3 Life Style Energy Efficiency

Figure 2-2 shows the amounts of annual resource use by humanity on the planet. We consume 7 gigatons of fossil fuels and 15 gigatons of oxygen to produce 22 gigatons of CO_2, 2 gigatons of food, 1.5 gigatons of forest, 1.3 gigatons of cement and 1 gigaton of steel.

The average human exhales one kg of CO_2 per day. Through fossil fuel consumption we emit about 10 times the amount of CO_2 per person compared to the amount emitted through breathing. We use fossil fuels to make our lives more convenient. Taken this way, each person on Earth uses the equivalent of 10 energy 'slaves.' But this is a global average; Japanese have the equivalent of 25, while Americans have equivalent of 55 energy slaves. At first glance the standards of living in all developed countries may seems similar, but seen through the window of energy consumption big differences exist. The issue here is one of lifestyles.

Figure 2-2. Amount of world resource usage and carbon dioxide emission.

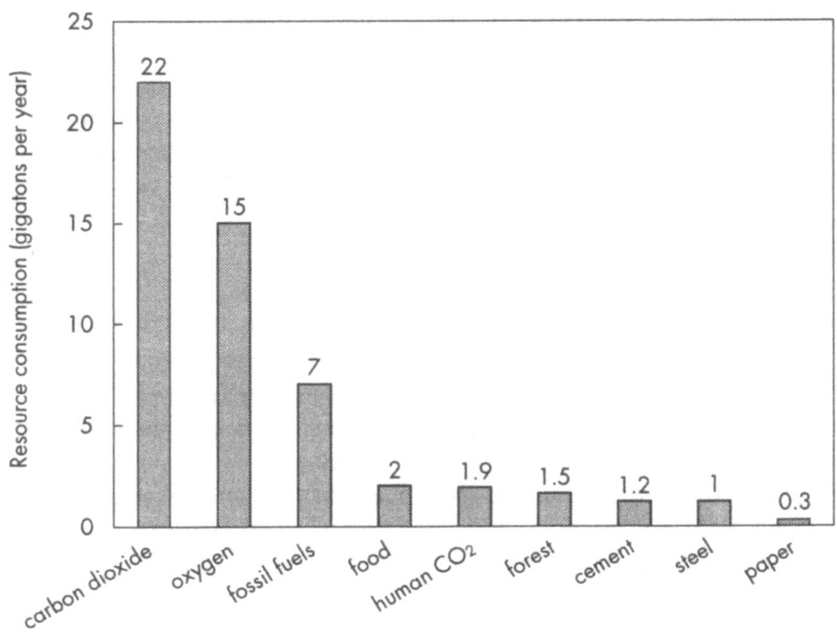

To take another example, today many electric appliances in households always have their main power switched on, ready for a signal from an infrared controller. It is as if we rub Aladdin's magic lamp, and the genie will appear. Electric ranges, video recorders, televisions, and audio devices are waiting to serve, not unlike Aladdin's genie. While on standby each device usually consumes between 1 and 10 watts of electricity. Multiplied by 8,760 hours in a year, the consumption becomes significant. Today it is common to have 50 watts-i.e. 10 to 15 percent of household electrical demand-generated from stand-by mode consumption. This could also be considered a lifestyle issue.

4. DECENTRALIZING ENERGY SUPPLIES

The energy supply system is currently in transition, moving from large scale centralized systems to a small scale decentralized systems. Reasons behind the transition are as described below.

4.1 Large Scale Centralized Energy Systems

Large-scale centralized energy systems have traditionally been built in the pursuit of economies of scale. Even if the capacity (i.e., production output per hour, or electrical output) of chemical plants or power plants were large, the cubic volume or weight of the plants, the land area, construction costs, and personnel costs would not increase proportionally. The reason for this is simple: given that the increase of capacity is cubed when the amount of material needed to construct the equipment is only squared, a smaller amount of materials is used per unit of increased capacity. These economies of scale reduce the unit cost of products or electricity generated.

It has been demonstrated, however, that economies of scale do not always work in this way. The large-scale plant requires a substantial amount of time to build. When the capital is unproductive for long periods of time, the capital performance is lower. Today the construction of a large power plant requires more than ten years. This long period is not suitable for an ever-changing dynamic society. When the construction starts on site, the situation changes, and the initial construction plans sometimes become meaningless.

It is widely known to the public that such a plant may have negative effects upon the local or neighboring community, such as the effects of air pollution and accidents which are often more serious than anticipated.

Another feature of large plants is transportation. The bulk transport of raw materials and fuels is necessary for the operation of the plant with the use of large tankers, and a port to reduce costs of volume transport, as well as an electrical transmission system. The locations of the raw material producers and end-users are generally some distance from the plant sites. Thus, transport issues may be seen as another challenge to traditional thinking about economies of scale.

4.2 Decentralized Energy Systems (DES)

Renewable energy sources, such as solar, wind, hydro and biomass, have five characteristics:[2] They

(a) use natural energy flows; they use the income of renewable energy, and do not deplete energy capital;
(b) are decentralized, available at the scale appropriate to the needs;
(c) are easy to comprehend, not a mystery to understand;
(d) are matched in scale and in geographical distribution to end-use needs; and
(e) are in an appropriate form for the end-use.

Decentralized energy systems (DES) consist of small-scale components, which in the past may have been considered economically inefficient from the perspective of the economies of scale. Other features of DES are as follows:

(a) the distance between supply and demand site is short;
(b) should problems arise, the potential for effects to the surrounding area are minimal;
(c) very little construction time is required, so capital is used efficiently;
(d) diffusion of the systems is rapid because they exist or can be installed in many places at once; and
(e) decisions about these systems can be made quickly, and innovations and improvements are easy to apply.

DES has some strengths in contrast to the weaknesses of centralized energy systems. However, even these strengths alone are not enough to be for them to substitute centralized energy systems with photovoltaic (PV) systems; a back-up system is necessary for operation on cloudy and rainy days. The back-up design could be in the form of electricity storage by battery for several days or direct connection to an electrical grid. If the user has access to an electrical grid connection nearby, then it would be cheaper to use the grid system, buying electricity from a grid while a PV does not generate electricity and selling electricity when a PV generates an excess of electricity. This particular design has become popular.

In this way, DES has been constructed using some parts of existing centralized energy systems. This may a peculiar situation only during the transition to DES, or on the other hand this kind of approach may always be necessary to some extent in the future.

4.3 Decentralized Energy Technologies

It becomes clear that the issue is the level of decentralized energy system technologies. They are constantly changing, and in a process of innovation. Here is a summary of current technologies.

4.3.1 Solar Heating

Solar heating is a direct, simple and practical use of solar energy. A domestic solar hot water heating unit consists of a 2 square meter black flat solar collector and a 200 liter water tank. This will supply a sufficient amount of 60 °C hot water for bathing. The use of solar heaters in households spread rapidly after the 1973 oil crisis and 5 million units are currently used in Japan. With a cost of nearly 300 thousand-yen per unit, it takes only about 7 to 8 years compared to the cost of using natural gas for heating. A solar system with a large collector can be used for drying and cleaning in industries that require temperatures up to 100 °C.

Other technology available includes systems to concentrate solar light using reflective surfaces or heliostats in order to generate high temperatures. The main target of experiments is to convert heat to electricity, but prototypes have not been successful in terms of economic performance. One of the merits of these systems is that if heat can be stored directly in heat-absorbing material, it is cheaper than storing energy in batteries.

4.3.2 Solar Houses

Solar houses are carefully designed with consideration of indoor heat flows, the capture of solar heat, and prevention of heat loss. The passive solar house is designed to acquire solar heat by generating airflow under the glass cover on the roof, in order to send warm air into an underground thermal storage facility, for later use for heating rooms during the winter season. From spring to autumn the hot air flow is used to warm water. The glass cover on the roof is used for radiation cooling in summer. Passive solar houses feature a simple design, which does not include a large pump and liquid-type solar collector. Several thousand passive solar houses have been built in Japan and the number is growing.

4.3.3 Photovoltaics (PV)

Photovoltaic systems convert solar light to electricity directly through the use of semiconductors. The conversion efficiency is 8 to 15 percent depending on the type. This rate is being improved and may be more than 20 percent in the future. The energy density of solar light is one kilowatt per square meter area and we can generate 100 watts of output from 1 square meter PV with a 10 percent rate of conversion efficiency. A 3 kilowatt PV system is enough to supply a sufficient amount of electricity for the demand of an ordinary household in Japan and requires only 30 square meters of roof area, which is easily available on an ordinary house.

A study of the potential for photovoltaics by the year 2015 in Japan found that about 10 gigawatts of electricity could be produced by installing PV systems on 500,000 single-family homes, as well as on housing complexes, schools, hospitals, post offices, government buildings, factories, warehouses, retail shops and hotels, etc. In addition, the potential of special electrical generation plants using PV could be 5 gigawatts, for a grand total of 15 gigawatts.[3]

If the costs decrease, and the number of homes installing these systems increases, the total electrical production using PV will exceed this estimate. In addition, lower costs would make it common to see any wall or roof that is well lit by sunlight to have a PV panel installed.

PV costs have already declined drastically, and continue to decrease. Today the production cost of PV is one million yen per kilowatt, giving an electricity price of 80 to 100 yen per kWh (assuming a 20 year lifespan of the system), which is nearly 3 to 5 times as much as the present electricity price from the electrical grid in Japan. Analysis of past cost reductions shows a learning curve such that for PV cost decreased by 80 percent with every doubling of cumulative production volume. As PV production is now on a learning curve we can expect significant cost reductions as PV production increases.[4]

4.3.4 Wind Power

Wind power has been used to produce electricity for years. Fiber reinforced plastics (FRP) have been used in the manufacture of modern propeller since the 1970s. Typically producing between 10 and 500 kilowatt of electricity, they are already considered economically viable. The global total electrical generation from wind power was nearly 7 gigawatts in 1997, including California, Holland, Denmark, Germany, and India, and a total 10 gigawatts worldwide was expected by the year 2000. The use of wind power has long been considered inappropriate for Japan because of the annual typhoons, but problems have been overcome and already 20 megawatts of generation capacity was installed by 1997. Because Japan has many good sites for wind power, such as coastlines, fields and mountain areas, wind powered electrical generation is likely to expand rapidly in the future.

4.3.5 Hydropower

Hydropower is also derived ultimately from the conversion of solar energy. It utilizes the power of water falling from dams in higher areas where rainwater has been collected. The global capacity of exploitable hydropower is immense. However, because large-scale hydropower dam construction causes serious impacts on the surrounding environment, small- and medium-scale hydropower is preferable to large projects.

4.3.6 Biomass

Biomass is the conversion of vegetation resources to be used as energy. Traditional uses of biomass include fuelwood and charcoal which have led to serious degradation of forest lands. However, it is possible to use biomass if planning includes systematic afforestation and the use of agricultural waste. Already methane gas is being generated the fermentation of waste. In addition, in Japan, electricity is being generated from burning household and retail store waste. In particular, this method of municipal waste disposal is growing today because of a shortage of final disposal sites for waste, and burning it reduces the volume of material. It is best to reuse and recycle first, but when this is not possible, incineration can produce electricity and the heat can be used to supply the demand for heat in the local area.

4.3.7 Cogeneration

Cogeneration is an energy system that is used to supply both electricity and heat by burning fuels. Power plants usually convert fuels to electricity with 30 to 40 percent efficiency, with the rest of the energy being wasted as low temperature heat. If this wasted heat is used as hot water for purposes of heating and drying, the total efficiency rises to 80 percent. It is necessary to determine the energy demand (electricity and heat), of such consumers as hotels, large shopping centers, and hospitals.

Today cogeneration relies on fossil fuels such as oil or gas, although it may be grouped along with DES as it is expected that cogeneration, will use biomass and fuel cells in the future.

4.3.8 Hydrogen

On earth, hydrogen is not directly available, however, it is often discussed in terms of its potential as a main energy source for the future. When burned, hydrogen emits only water; no other pollutants are released into the environment. Currently, hydrogen is produced by water electrolysis or decomposition of methane gas or alcohol. One of the possibilities of hydrogen use is the fuel cell. The fuel cell converts hydrogen to electricity through a silent electrochemical reaction, without combustion. A new trend is to use hydrogen for fuel cells to power automobiles. In 1998, Ballard Power Systems Inc., a Canadian company, developed a small scale, lightweight fuel cell with a proton exchange membrane (PEM), weighing only 60 kg and generating 50 kilowatts of electricity, similar to a small gasoline engine. Daimler Benz invested in Ballard Power Systems Inc. and plans to produce 40 thousand automobiles per year equipped with fuel cells in 2004. Gasoline or alcohol is used on board as fuel and electricity is generated by fuel cells that drive the wheels. The fuel cell car is expected to drive future competition in the automobile industry.

Hydrogen will be widely used in the future. It will be produced by electricity from PV. In 1998, the National Renewable Energy Laboratory (NREL) in the United States succeeded in doubling the efficiency (to 12.4 percent) of converting solar light to hydrogen by developing a new device equipped with the functions of both PV and electrolysis in one chip. The U.S. Department of Energy forecasts that hydrogen will supply energy to 4 million households in 2010, and 10 million in 2030. It is difficult to store electricity, but hydrogen can be stored effectively. Thus, it is likely that the future cost of hydrogen will be only slightly higher than the cost of electricity.

4.4 Efficiency of Decentralized Energy System

A decentralized energy system is located near a demand site for purposes of minimizing the loss of energy in transport, space requirements and conversion losses. Even for these reasons alone, DES is expected to be relatively efficient. One good example is cogeneration. Cogeneration systems are inherently decentralized because they are typically located close to the end-use demand in order to supply electricity and heat. From a physics perspective cogeneration employs a 'cascade' utilization of thermal energy: high temperature heat is used its mechanical power or electricity generation and low temperature heat for drying, heating and hot water. Though energy conversion efficiency is 30 to 40 percent in the case of electricity only, it goes up to nearly 80 percent by cogeneration.

Through PV based decentralized energy systems it is not difficult to supply all of a household's electricity demand. If it is installed in a house having a south-oriented roof, a 30 square meter PV can generate 3 kilowatts electricity. Connection to the electrical grid system may be basically unnecessary in this case, but since the generated electricity is of larger quantity than the demand in the house, it is possible to send it to the grid system. Storing excess electricity in a battery for several days enables it to be supplied on cloudy and rainy days. If the house is connected with a grid, then without the need for an expensive battery, it is possible to send excess electricity to a grid. Other innovations of battery and fuel cells are under way, which may boost the capacity to store excess electricity in the home.

The energy loss through transmission and distribution in grid systems is now 8 percent. In case of pumped hydropower the number is larger. From the perspective of thermodynamics, running an electric heater using electricity from the grid is the least efficient way of using electricity. Solar heat and solar houses are impossible in centralized energy systems, since the transport of heat is limited to short distances. District heating systems can only carry heat to a limited area.

5. DES AND THE ECONOMY

The relation between decentralized energy systems and the economy is an important issue. This section will discuss the transition to DES, costs associated with it, stages of development, and policies to promote it.

5.1 Transition to DES

If a smooth transition to decentralized energy systems is not made, humanity will find itself even more dependent on centralized energy systems in the future. This would result in the increase of fossil fuel consumption and lower energy efficiency. If nuclear energy use were expanded, we would have the greater problems associated with nuclear proliferation, plant safety and waste disposal. If the use of coal were expanded, air pollution and global warming would worsen. It is clear that increasing dependence on centralized energy systems would some day lead humanity to a dead end.

One acceptable fossil fuel to use is natural gas. Because natural gas has a lower carbon content than other fossil fuels, the CO_2 emitted per unit of energy produced is equivalent to only 60 percent compared to coal and 80 percent compared to oil. The safety problems of large-scale energy system remain however, with giant tanks needed to store gas and liquefied natural gas (LNG) tankers that would be required for transport. On the other hand, natural gas can be used to drive gas turbines in cogeneration systems. In addition, the pipelines used for natural gas transport today could be used for hydrogen transport in the future if we develop hydrogen-based economies. In this sense, the use of natural gas, could promote a transition to decentralized energy systems in the future.

The introduction of decentralized energy systems that use renewable energy is technically feasible today. However, they are often not yet economically competitive with fossil fuels. In many cases this economic disadvantage will be overcome as the technology moves along the learning curve with broader use.

Much of the energy demand in the household and commercial sectors is relatively small in scale and thus suitable to be met with DES. In the industrial sector however, some large-scale energy uses are inappropriate for DES. Examples include iron and steel, cement, paper and pulp industries. For these industries, fossil fuels will probably be needed for some time in the future. Gasoline or portable liquid fuels have high energy density per unit of volume or weight and are therefore suitable to satisfy the energy demand of automobiles in the transport sector. For this reason, the hybrid car is more practical than the electric car that carries a battery weighing nearly half the car's weight. With that in mind, it is possible that decentralized energy systems using biomass could supply liquid fuels to automobiles in the future.

5.2 Cost of DES

Energy systems supply the basic driving force of the economy. If energy costs are low, the economy overall will have a lower burden of energy expenditures and it is possible that this would accelerate the economy. This section will discuss this possibility.

Even in conventional energy systems, if oil prices go up, the economy slows down. If the price of renewable energy remains high, this too will slow the economy. However, if global warming becomes a reality, the consequences are much greater than just a slower economy. Indeed, if we were able to continue using conventional systems in a stable way over the long term without causing any problems in the future, new energy systems would not be necessary.

However, in the long term the costs of decentralized energy systems tend to decrease; in particular, this is true with renewable energy, because operating costs are inherently low. In the long run investments in these systems bring excellent returns. In addition, it is essential to recognize their value as measures to mitigate global warming.

There are two ways to speak about future market prospects for a technology or product. For example, one way goes like this: "There is a market of 2 trillion yen and 300,000 jobs in the construction of high speed, multimedia fiber optics systems." Another approach goes like this: "Investment in decentralized energy systems cannot compete with fossil fuels because they will cost 2 trillion yen and the costs of energy they produce are still expensive." The difference in thinking is whether one recognizes the existence of a market, or whether one sees only the expenditures needed. In reality, the housing industry in Japan is already using "energy independence" as a tool in the marketing of new homes. The PV market in Japan is already as big as 100 billion yen.

Future economic planning should take into account the perspective that new energy systems will create real demand, sell products and create employment. From this viewpoint, decentralized energy systems may have greater potential to benefit the overall economy than centralized energy systems that depend on fossil fuels. Even if the costs are high, if society needs something, a market will naturally emerge for it.

5.3 DES and Stages of Economic Development

The traditional assumption has existed that the decentralization of energy systems happens only after countries reach a certain economic level. The thinking generally is that mainly in developed countries, while addressing the problems of centralized energy, the advantages of decentralized energy have emerged. Or the technological advances in those countries made decentralization possible. So in those countries it was after completion of the basic infrastructure of society-roads, railways, bridges and cities, etc.-when the energy demand for material resources started to decline that decentralization begins.

There are, however, other possible scenarios. One example is the rapid penetration of cellular phones, now proceeding rapidly in Southeast Asian countries. This allows communication without telephone wires. Similar patterns will occur in energy fields. When mass production of PV arrives and costs of installing photovoltaic systems in developed countries go down, it is predictable that developing countries will want to build decentralized PV systems rather than large centralized energy systems. This type of development is known as "Leap Frogging." Developing countries need not follow the same paths as developed countries. They could jump forward using the advanced technologies developed in industrialized countries.

5.4 How to Promote Decentralized Energy Systems

The cooperative efforts of inventors, engineers, industry, banks, local authorities and national governments are essential in order to spread the use of decentralized energy systems in society.

The Japanese government provides subsidies to individuals who install photovoltaic systems in their houses. At the beginning the subsidy was one-half of installation costs but today it is one-third. At current rates, this means that such households produce electricity at 90 yen per kWh, compared to about 23-25 yen per kWh when buying from the electrical grid. The success of the program is relying on the volunteer spirit of these users; otherwise we must wait until mass production brings down the costs of PV systems.[5]

In contrast, the town of Aachen in Germany added 1 percent to the price of electricity in order to buy the power produced by PV and wind turbines. The system is set up so that people who have installed PV or wind turbines at an early stage do not suffer financial losses. This system is now spreading to other towns in Germany.

Now would be a suitable time for governments to introduce what could be called a Solar Energy Supply Act which sets targets every few years for the spread of photovoltaic and wind power systems. Central and local governments should set certain targets for the proportion of their own electricity purchases that comprise renewable energy (e.g., 1%, 2%, 5%, etc.). Electrical utility companies should similarly set targets for the proportion of the electricity they supply from renewable energy. Further, consumers should be able to choose the proportion of electricity they use that comes from renewable energy. For example, if they choose fifty percent, they should be willing to pay a little more than otherwise.

Now would also be a suitable time to introduce laws that require improvements in energy efficiency. Electrical appliances that operate with stand-by-mode electrical consumption could be eventually banned (initially, standards could be set at 1 milliwatt or less). Tax incentives could be given to promote the use of hybrid cars (these are already in place for companies and local governments in Japan). Likewise tax incentives could be offered for efficient electrical appliances, tax exemptions for people who install renewable energy systems and incentives to promote cogeneration systems. It is important to encourage innovation of our social institutions to foster changes toward new energy systems.

6. FROM ENERGY HUNTING TO ENERGY CULTIVATING CIVILIZATION

As described above, in the twenty-first century a transition will likely occur from the current energy systems to more efficient decentralized systems that use renewable energy sources.

Geophysicist King Hubbert, who was the first to point out the finiteness of fossil fuel resources said that if one were to look back over history, our age of dependency on fossil fuels would only be a needle's width on the clock of time. Clearly the width of that needle spans the twentieth and twenty-first centuries.

Looking back with a long-term perspective, we can recognize the experience of humanity in the transition toward food production. Initially, man relied on hunting animals for food. This could be called a "Hunting Civilization." After being hunted for millennia the numbers of animals decreased. In need of other food sources, humanity turned to farming and created what is called "Cultivating Civilization." The present situation surrounding energy and the environment are very similar to this historical transition. Mankind is now at the end of the "Energy Hunting Civilization" which involves extracting fossil fuels from the ground. The direction in which we are moving will probably be something parallel to what happened with agriculture. Our future civilization may therefore be referred to as an " Energy Cultivating Civilization."[6] This transition will require time, but it remains to be seen whether humanity can succeed in this transition or not. The big question is whether humanity will adapt before the finite supplies of fossil fuels run out. The significance of New Development Patterns of society is about this very challenge of transforming ourselves into an Energy Cultivating Civilization.

7. CONCLUSION

This paper covered New Development Patterns and energy. The societies based on New Development Patterns must aim toward forms of development that minimize the effects of energy consumption on the environment. The new patterns must also move humanity toward energy strategies that will solve the problem of global warming.

In New Development Patterns, the first priority is to improve energy efficiency. Unless the challenges of efficiency are dealt with, there will be no other progress.

Already many economically viable, energy efficient technologies exist that could be widely used if even small subsidies and more information were provided to promote their utilization. Next, through the use of renewable energies, the supply of energy must be decentralized. These renewables are very compatible with the overall direction society must move. Technologies for decentralized renewable energy are already well-known and are being applied. However, a little more progress is needed in terms of economic viability for them to compete with conventional energy sources. There are vast opportunities for the diffusion of renewable energies if central and local governments create effective policies, and if industries refine their marketing strategies. As the use of renewable energies expands, costs will decrease enough to make them more viable.

Considering history with a grand perspective, the period from the twentieth to the twenty-first century will have been a transition period from "Energy Hunting" to "Energy Cultivating" civilizations. The transition has only just begun.

REFERENCES

[1] A.K.N.Reddy, R.Williams, and T.B.Johansson, *Energy After Rio*, Prospects and Challenges, UNDP, 1997.

[2] A. Lovins, *Soft Energy Paths, Toward a Durable Peace*, Friends of Earth, USA, 1977.

[3] J. Minakami (Energy Statistics at MITI), Manual of Photovoltaics Introduction, Eimuzu Publishing Co. 1995.

[4] H. Tsuchiya, "CO_2 reduction scenario for Japan", *Nikkei Science*, November 1990.

[5] S. Kurihara, *Environmental Citizen Revolution*, Energy Conservation Center, Japan, 1998.

[6] H. Tsuchiya, *Energy Cultivating Civilization*, Toyo Keizai Shinposha, 1980.

Chapter 3

ENERGY AND THE NEW ECONOMY

HARUKI TSUCHIYA
Research Institute for Systems Technology

1. INTRODUCTION

Innovation is a key driving force in modern society. Microelectronic technologies provided tremendous innovation at the close of the twentieth century. The technical changes of computer hardware and software have rapidly penetrated into societies. In the United States it is commonly thought that the main thrust of recent economic growth was powered by information technology (IT).

This economy driven by IT has been called the New Economy. IT is being applied and used in diverse areas of economic activity, and has been credited with having improved the profit performance of many corporations. The wide use of IT has promoted a transition toward information-based societies. Some key outcomes of IT include the rapid expansion of the use of the Internet, ubiquitous use of electronic mail, an increase in electronic publishing, the emergence of electronic commerce, and experimentation with electronic money.

It seems inevitable that IT will permeate entire economies. Microelectronic innovation will stimulate new progress, as well as potential problems. On the other hand, IT can play an important role in the twenty-first century to improve the efficiency of energy and resources, and to reduce carbon dioxide emissions in order to mitigate global warming. IT affords the possibility to improve the efficiency of society in terms of transportation and the use of paper.

When we think about new development patterns, it is important to consider the effects of the New Economy upon society and its consumption of energy and resources, an analysis of which will follow. As shown in the next section, in the future energy systems may be more decentralized and based on renewable energy. This paper will also address the relationship between the New Economy and the decentralization of energy systems.

K. Matsushita (ed.), Environment in the 21st Century and New Development Patterns, 69–85.

2. THE NEW ECONOMY

There have been several stages of transition toward the New Economy, of which IT is a major element. At the first stage, IT contributes to improvements in the performance of the economy as may be seen today in the United States. At the second stage, many IT-based industries emerge and become the new driving forces of economic growth. This process is now underway. Material production has reached a level of maturity in developed countries to the extent that further economic growth will depend upon increases in non-material based sectors. These will depend less on steel, paper, cement and other materials to produce what people consider useful. They should benefit societies, create jobs and contribute to income redistribution. Is this possible in software and information industries? This is not a simple question.

The knowledge-based industries include mass communication, publishing, and education related industries. The entertainment-based industries include television, movies, and computer games. The software industries develop industrial information systems. Others are industries related to sightseeing, travel, leisure, sports and arts. The main content of these industries is information and knowledge, and because of that they will use IT. In the New Economy it may be possible to shift from the energy-intensive economies of the past which required massive amounts of material resources, to information-intensive economies.

IT networks are growing beyond national borders to connect on a global scale. A new world is emerging in which companies in rural towns of Japan are expanding their business with the whole world; companies in small villages in South America can deal directly with large companies in developed countries. The local and the global are being connected. As this happens, the roles of nations will become relatively less important. In order for small communities to begin to connect with the world in its entirety, it is important that governments do not interfere excessively and that local governments support these new trends.

3. ASPECTS OF THE NEW ECONOMY

The New Economy is expected to promote new businesses based on IT communication systems, such as those offering transportation alternatives, electronic alternatives to paper, faster communications and computer processing and more efficient use of energy and resources. Corporate activities will involve less waste thanks to better use of information: instead of moving people around, meetings and negotiations can be conducted using IT. The management, logistics and purchasing departments in companies will be organized based on IT systems.

Although to date no comprehensive studies have looked at the effects of the New Economy on resources and energy, it is possible that their demand will change drastically. Here we will discuss this possibility.[1]

3.1 Home Offices

The term SOHO (Small Office, Home Office) refers to working at a small or home office, using computer networks. In this style of work, also known as "teleworking," communication systems play a central role. It is becoming part of a new lifestyle and is popular among people who are intellectually independent. Their outputs are not hardware products, but rather, are exemplified by computer software, reports, proposals, statistical survey reports and manuscript articles for magazines.

There are many types of work that no longer require people to meet at the same place. For this work the creative driving force is the ability to freely access new information and ideas. Those who work in such style need only a computer and a communication line. Information is collected through the Internet and e-mail, meetings are conducted 'on-line' with clients, and the finished product is delivered electronically as well.

"Telecommuting" is a work style, which involves the employee working from home without the need to commute or the necessity of having a desk at the company office. Telecommuters are employed by companies, but must manage their time on their own.

The costs associated with television conferencing systems have been reduced by using personal computer systems, small television cameras and telephone lines. Telephone lines used for establishing Internet connections were quite slow, generating only poor quality pictures, and thus inadequate for sending thirty images per second, as in normal television. However, optical fiber lines are substituting conventional telephone lines, and "Internet 2" is likely to perform 1000 times faster than older systems. By connecting many cities with one universal network, many people will be able to work together simultaneously with the aid of television conferencing systems, and without the need of transportation.

3.2 Intelligent Transportation Systems

Intelligent transportation systems, which involve the placement of receivers on vehicle and roadside information supply systems, are being developed to aid drivers through or around traffic congestion to destinations without experiencing loss of time. These systems are being introduced on highways and roads in urban areas. More than 2.6 million cars have been already equipped with navigation systems, which allow drivers ease in planning ways through complicated routes in congested areas. The use of such navigation systems is supposed to improve driving speed and fuel economy. Furthermore, the information systems are designed to supply data pertaining to shopping and restaurants to the receivers in automobiles. However, some critics counter that even if traffic congestion is decreased, other drivers and cars will be attracted to the roads, reversing and progress in reducing congestion. At this point it is still difficult to predict the net outcomes of these systems.

3.3 Electronic Commerce

The Internet promotes changes in distribution systems. Books, cars, stocks and real estate are some items already being sold through electronic commerce. One of the most successful venture businesses in electronic commerce was Amazon.com, which began by selling books through the Internet. Consumer as well as industrial markets have similar electronic commerce potential for purchasing materials and parts from companies, who are able to purchase materials for the lowest cost and are able to accept bids through electronic networks. Electronic commerce has the potential to link companies on a global scale, even those that have never before done business together.

A new method known as "one-to-one marketing" is designed to utilize the detailed consumer preference information contained in database systems including brand consciousness, luxury preference, price sensitivity, whether they go skiing in winter and visit resorts in summer, etc. Databases enable companies to predict behavior and pinpoint suitable information about the goods that will attract individual consumers. IT offers appealing methods for marketers that differ from conventional mass marketing methods based on television and newspapers that are not as efficient at targeting consumers.

Individual production systems are being developed in which a product is produced after an order has been received through electronic commerce and sent directly to the customer. As a result, the necessity for shops and warehouses will be either eliminated or reduced. Production based on sales forecasts and the keeping of excess inventories will become less necessary and stocks of unsold goods will be reduced. Virtual shopping and virtual factories will transform distribution systems.

Electronic money also affects trends of electronic commerce. It is a new form of exchange, functioning only in networks, invented as a substitute for credit cards, which can move freely beyond national borders. As a result of IT, economies are accelerating as well as saving time resources.

On the other hand, electronic commerce will inevitably increase the amount of electronic crime through electronic networks. Many people assume that criminal activity will be suppressed by encryption systems, but new problems posed by electronic hackers will occur after electronic commerce has grown further. The eventual hazards of electronic crime are not yet clear at current scale of electronic commerce.

3.4 Electronic Books and Newspapers

It has become common to view text and images stored in CD-ROMs on computer screens. A number of developments have the potential to dramatically alter the way people read.

Network magazine Wired (July 1998) introduced prototypes of three types of electronic books in the United States. These electronic books are handy book-shaped liquid crystal display devices. They have memories to store text and images. A reader needs only to push a button in order to turn the pages. One can read from many different books simply by downloading the contents of a new book into the device's memory.

The Softbook was designed to simulate the experience of a conventional book, with a leather cover and one page of the book displayed by liquid crystal display. The Everybook was designed to display two pages in the same manner as conventional books by two liquid crystal displays. Both are connected by modems to networks, allowing readers to download the contents of books via telephone lines. The Rocketbook is designed to have one page displayed, with the contents downloaded via a personal computer.

Electronic books are free from some problems of conventional books, such as going out of print and returns. Currently an average of 30 to 40 percent of books and magazines are either returned to the publisher from the retailer or unsold. Bookstores are often faced with the challenge of having to return books and magazines. There are simply too many books published on any given day throughout the year. Returned magazines are disposed of, as there is no value in out of date material, creating a significant waste of paper resources. Should conventional books, magazines and newpapers be substituted by electronic books, such problems would not exist, and readers would be able to access a variety of books at any time from the electronic libraries.

In Japan Sony and NEC have conducted trials during the last ten years. The Data Discman from Sony consists of a small display and an 8 cm compact disc (CD-ROM) drive, which contains an encyclopedia or dictionary . NEC's The Digital book is a reading device with a liquid crystal display and floppy disk drive, which contains essays, comics and novels. These trials could have yet to reach the mass market, but the companies aim to revolutionize books for the first time since Gutenberg's invention of the printing press.

Japan's top publishers created the Electronic Book Consortium in the autumn of 1998, with a plan to create a system to transmit the contents of books through communication satellites to automatic book selling machines at convenience stores and kiosks in railway stations. Consumers will be able to enjoy the books in their reading devices , which simulate a book with left and right pages.

Electronic books will contribute to a decrease in the waste of paper, and improve the value of books and newspapers. Newspaper companies already utilize the Internet to supply news stories. This has lessened the need for newspapers to be hand delivered to homes and offices on a daily basis. In other new products, pager-like devices receive fresh news items and will fit into a person's pocket.

Problems that still require attention are "readability" or user-friendliness of the liquid crystal displays, the weight of the devices, and the lifetime of the battery.

3.5 Electronic Mail

Some people credit e-mail for decreasing the amount of time spent on air travel for business meetings. Files can be exchanged and edited online, resulting in increased efficiency and better finished products. This is a completely new business style. E-mail, unlike telephones, does not have to interrupt one's work . An increasing number of people are using email at the rate of 100 e-mail transactions per day.

3.6 Precise Computer Control

Production processes controlled by computers have improved the quality of products, and contributed to a decrease in the loss or rejection of materials . Furthermore, IT could stimulate design processes such as recycle design, environmentally benign design and design based on life cycle assessments. These new design processes would be impossible without IT. Materials-recycling requires computer management during the production stage, using materials that are easy to recycle and designs that allow easy dismantling.

3.7 Energy Star Project

Personal computers in offices are often left with the power on even while not in use, posing a significant energy consumption problem. To solve this problem, the Environment Protection Agency in the United States developed the "Energy Star" project, which has also been introduced in Japan. To qualify under the standards of the project, personal computers with liquid crystal displays and their peripheral devices are required to operate at fewer than 30 watts of electricity consumption. The Energy Star project has since been applied to other office equipment and has penetrated the market, with electric appliance manufacturers labeling their products with the Energy Star mark. The regulation limit should be decreased on a gradual basis to account for corresponding technological innovations. As the available number of IT devices increases, it remains important to regulate energy consumption.

4. EFFECTS OF THE NEW ECONOMY ON RESOURCES AND ENERGY

An analysis of the effects of IT in the New Economy, with an emphasis on the effects upon energy and resources will follow. The following discussion will consist of a quantitative analysis and reference to potential problems caused by IT.[2]

4.1 Provision of Information on Resources, Energy and Environment

In the New Economy the value of information is expected to increase and anyone will be able to freely access a variety of information. The Internet enables the transmission of information on resources, energy and environment to a vast audience.

The speed with which the Berlin Wall collapsed surprised many observers. People in Eastern Europe had been obtaining information through television, movies and music from the democratic world. During the last ten years of the twentieth century the Cold War came to an end. Many people believe that the media was a major factor in these historic changes.

A similar phenomenon may occur with respect to information on resources and the environment. People can see the results of observations about the global environment: measurements of CO_2 in many places in the world, temperature increases in big cities, melting of the permafrost in Alaska, retreating coastlines worldwide, rising sea level in small islands in the Pacific Ocean, and damage done to forests by acid rain. This awareness could turn people's attention towards solutions of the problems.

4.2 Paper Alternatives

A 1996 report entitled "Savings of Energy and Resources in Books, Magazine and Newspaper" by CRIEPI (Central Research Institute for Electric Power Industry), Japan, predicted that the trend towards electronic handling of information would significantly affect books, magazines and newspapers. It forecasts that by 2010, 30 percent of these publications would be electronic in a normal scenario and 50 percent under an accelerated scenario. The majority of predicted energy savings come from reductions in the use of material inputs and production of paper, and partly by a reduction in transportation costs.

4.3 Electronic Books and Newspapers

It is expected that newspapers, books, and magazines which are read only once will be replaced by electronic publishing, given environmental concerns around energy and resource consumption. It is possible to receive the information of edited text

Table 3-1. Electronic Publishing and Energy Saving.

Scenario	Annual Total Energy Saving (10^{12} kcal)
Normal Case(30% electronic)	12.1
Accelerated Case(50% electronic)	20.6

and pictures carried by telephone lines and radio waves, on small handy displays in homes and offices. The reader needs only to push a button in order to turn pages as one would do when reading a conventional book. The information is stored in flash memories holding several megabytes of data and displayed on low electricity consuming, reflection-type liquid crystal displays.

This electronic reading device would weigh 300 grams and consume 2 watts. For estimation purposes it is assumed that under normal circumstances the device would be used one hour a day for three years and the battery would be changed three times. Based on this it is possible to calculate the energy embedded in the display devices and battery, as well as the energy needed for displaying contents. The results are shown in Table 3-2, based on downloads of one megabyte of information over two minutes via telephone lines.[3]

Average time spent for reading newspapers is 20 minutes per person. Assume three people read newspaper hence spending one hour per paper per day, and eight hours to read one book. The amount of energy put into materials such as metal and plastics is 1 to 3 x 104 kcal/kg. If we assume the energy required for manufacturing one electronic reading device to be 30,000 kcal/kg, it turns out to be 6.31 kcal per day. As shown in Table 3-2, these electronic devises bear possibilities to conserve papers, and to increase energy efficiency by 40 to 220 times. However, there are also disadvantages. The whole page cannot be looked at a glance on the electronic device as in the conventional newspaper, and from a technical side, there are problems to be improved such as the effect of the battery to the environment, and the effect of the liquid crystal to eyesight.

There is also a possibility of newspapers transmitted through radio waves. The information is transmitted through radio waves used such as for pocket bells, and it is received by a small receiver to be shown on a display. Compared to those through communication network, use of radio waves allow large decrease in energy consumption when used by certain number of users, and is also more efficient than using cable communications.

Table 3-2. Electronic Newspapers and Books.

Energy Consumption	Newspapers	Books
Embedded in Electronic Reading Device (per day)	6.31 kcal	6.31 kcal
Data Transmission	14.85 kcal	14.85 kcal
Display of Data	4.5 kcal	18 kcal
Total for Electronic Publication (a)	25.65 kcal	39.15 kcal
Total for Paper Publication (b)	1,228 kcal	8,674 kcal
Energy Ratio (b/a)	47 times	221 times

4.4 Satellite Office and TV Conferencing

The Japanese Ministry of Posts and Telecommunications estimated in its 1993 report "Contribution to Environment" that in the beginning of twenty-first century decentralization of offices and television conference systems would change business practices, increase telecommuting, decrease the amount of transportation and save energy, as shown in Table 3-3.

Low-cost television conferencing systems using personal computers have become quite common, and the number of people in Japan already teleworking in some way was estimated to already be 950,000.

4.5 E-mail, Fax and Postal mail

Fax transmissions are thought to be more energy efficient than regular postal mail. Electronic mail is credited as even more efficient, given that no paper is required. This is an issue of substitution between communication and transportation. Of course it would incomplete to consider only energy consumption; other advantages of faxes over postal mail include speed and simultaneous distribution.

Table 3-3. Reduction of Trip Demand by Satellite Offices and TV Conference Systems

	Satellite offices	TV Conference Systems
Automobile*	1,008	8,942
Bus*	73.8	
Railways*	10,842	
Private Railways*		437
Air*		4,496
Total CO_2 Reduction	107,850 tonnes C per year	217,772 tonnes C per year

*Units: million person-kilometers per year in 2010.

The energy consumption related to the transmission of information using telephone lines is illustrated in Table 3-4. A person who picks up a telephone is connected with a 958 watt electrical load. [4,5] A medium-sized fax machine sends one A4-size sheet of paper in 12 seconds, with an electricity consumption of 40 watts, but it consumes 15 watts while operating in standby-mode. It is not unusual for a fax machine to remain on standby for an entire day; it is the standby mode, which accounts for nearly all the energy demand of a fax machine. Assuming a fax machine sends and receives 50 transactions per day, and counting 30 seconds of use of the telephone line each time, the energy for sending a page by fax is calculated as shown in Table 3-6. The number of letters sent by post per year in Japan is 22.3 billion as shown in Table 3-5, in which the transportation energy of postal mail is calculated by using an estimate of energy consumption at the post office combined with the cost of transportation.

The results of the calculation depend upon the frequency of the use of the fax machine; for example, the fax is nine times more efficient than postal mail if used 50 times per day. Alternatively, fax is only 1.8 times more efficient than postal mail when used only 5 times per day. Of course in the case of sending long documents, the relative efficiency of regular postal mail improves. The calculation did not allow for the inclusion of any envelopes required for the mail, or thermal fax paper on the receiving end of the fax transmission. It may be noted from this analysis that if the standby electricity consumption could be decreased, faxes will become significantly more energy efficient.

Table 3-4. Demand and Energy Consumption of Telephones (1993).

	Demand and Energy consumption of Telephones
Number of Telephones	55,830 thousand
Total Number of Calls per year	73.5 billion
Total Hours of Call per year	3.444 billion hours
Number of Calls per Telephone	1316.5 times per year
Average Time of Each Call	2 minutes and 49 seconds
Electricity Consumption of NTT	3.3 billion kWh per year
Electricity Consumption of one Call	0.0449kWh (101.0 kcal)
Electricity Consumption of Call per hour	0.958 kWh/h

Table 3-5. Demand and Energy Consumption of Regular Postal Mail.

	Demand and Energy Consumption of Postal Mail
Number of Items Total	22.3 billion per year
Number of 1st Class Mail (Envelope)	10.7 billion per year
Number of 2nd Class Mail (Post Card)	5.8 billion per year
Electricity Cost at Post Office(24,107 offices)	21.3 billion Yen per year
Transport Cost	233.4 billion yen per year
Energy Demand at Post Office per Item	50.6 kcal per item
Energy Demand at Collection and Distribution	24.9 kcal per item
Energy Demand of Long Distance Transport	269.0 kcal per item
Energy Demand	344.5 kcal per item

Note: Energy demand is author's estimate.

Table 3-6. Energy Consumption of Fax Machine.

Capacity	Average Time of Use	Average Energy Demand Per Use
Medium Sized Fax Machine (40 W) Standby mode (15 W)	30 seconds	19.2 kcal when 50 uses per day 192 kcal when 5 uses per day 384 kcal when 2 uses per day

Extending this estimate to e-mail, a notebook-size personal computer consumes electricity at a similar rate to a fax machine. By contrast, a desk top personal computer using a cathode ray tube (CRT) monitor consumes five to ten times more electricity than a fax machine. Service providers, which operate 24 hours a day, should be counted in the case of e-mail. One provider serves several hundreds of personal computers , resulting in electrical consumption that is relatively small by comparison to the total consumption by mainly end users, i.e., smaller computers. As a result, total electricity consumption would not depend on the number of e-mail transactions, but rather on how long personal computers are switched on. Thus, e-mail transactions sent by personal computers with liquid crystal display are more efficient than fax. When used only two to three hours a day, e-mail transactions by CRT type personal computers have nearly the same efficiency as fax, however, if powered on for an entire day, they are less efficient than fax.

4.6 Decrease of Transport Loss

The application of IT to transportation management has enabled companies to plan precise delivery schedules. Such schedules are designed to decrease the number of empty trucks in operation through IT-based truck allocation planning, in order to carry consolidated loads of cargo to neighboring destinations. This would improve efficiency by twenty percent. Furthermore, the installation of mobile computers which are used to send and receive information in order to confirm the destinations with a control center have been installed in vehicles, enabling drivers to maximize productivity. IT will promote the saving of energy, while enabling the efficient management of transport.

4.7 Increase of Small Packet Transport

IT has accelerated information transfer, but IT is limited to the handling of information. Of course, since it is not possible to transfer real goods and materials through IT, the use of the conventional transport system will continue. Electronic commerce through the Internet enables an individual buyer and seller to conduct business transactions. Increasingly, small, packaged products will be delivered directly from the factory to the individual buyer who performed the simple task of making an order by computer. This type of business will require an alternative means of transport, such as small or individual package transport.

Small package transport may or may not contribute to higher total energy consumption than bulk type massive transport, but the amount energy consumption associated with shopping by car will be decreased. Generally the distance of transport of goods will be shorter in the case of small package direct transport than in the case of bulk transport. Under some circumstances, Internet shopping is considered to be more energy efficient

The conventional transport system, which carries goods to local transfer points through bulk cargo delivery and then converts goods to small package, was once thought to be an efficient system. Research has indicated, however, that the current system requires substantial improvements. It remains unknown as to whether the energy consumption associated with transport will increase or decrease as a result of these changes in transport.

4.8 Energy Efficiency Improvement of Distribution System

As electronic commerce becomes increasingly popular, the distribution systems will require alterations. The energy inefficiency of conventional shops that wait for customers will become evident. Shops may one day become unnecessary, as "one-to-one" production systems are developed to produce goods after receiving individual orders. Eliminating the need for shops and warehouses will diminish energy consumption required for their heating, cooling and lighting. Given that the amount of lighting in shops is often more than needed to display items for sale, the decrease in energy consumption could be enormous. Consumers using electronic commerce would simultaneously reduce their own time used in shopping and save the energy required for transportation.

The use of electronic commerce within industry is also important. Internet business allows companies to discover new partners, without depending on conventional business relationships. Negotiation of prices, quality, specifications and deadlines may be conducted more efficiently via the Internet. Sales divisions will find markets worldwide and target new and larger markets. As mentioned above, transportation for meetings may be reduced and negotiations may occur more efficiently through IT.

4.9 Efficient Products with Microelectronics Control

An increasing number of products on the market are equipped with embedded microelectronic devices. One example is the computer software contained in new hybrid cars, which run on electricity and gasoline. New applications required new, high technology software for engines, motors, generators, batteries and drive shafts to transfer power smoothly and maintain optimum driving conditions. When a driver pushes the brake pedal, software controls the braking power, while a portion of the energy is converted to electricity through the use of generators. Software also determines how much engine power is converted to electricity for the battery or transmitted to the drive shaft. As a balance is required, a complex calculation is necessary. As shown in this sample, the conventional industries are also developing procedures whereby software would be embedded into products in order to facilitate improved performance with low cost IT. This is yet another new trend in the application of IT.

4.10 Standby Electrical Consumption

A recent survey measured standby electrical consumption in households at around 50 watts, accounting for 10 to 15 percent of the total household electricity demand. In homes many electrical appliances have no main switch to turn off the power consumption, such as microwave ovens, remote control air conditioners and electric kettles which maintain a constant water temperature even when not in use. One cannot switch off such appliances unless the plug is pulled from the electrical outlet in the wall.

Functions such as the clock display on household electrical appliances and television preheating for instant startup are powered on a continual basis, consuming large amounts of energy without being noticed by users. Electrical appliances such as television, video, audio appliances, fax, and phones with high standby energy consumption should be prohibited from manufacture and be replaced with those

Table 3-7. Examples of Standby Electrical Consumption of Appliances.

Electrical Appliances	Standby Electrical Consumption (W)	Annual Electricity Cost (U.S.$)
TV	1.9	3.4
Video Recorder	8.2	15.2
Audio Stereo	11.0	20.4
Air Conditioner	7.3	13.5
Telephone(Parent)	4.7	8.7
Telephone(Child)	2.2	4.1
Microwave Range	5.5	12.4
Wash Toilet	6.0	10.8
Electric Tooth Brush	4.5	8.1
Total	51.3	88.5

Source: Jyukankyo Research Institute, Tokyo, 1997

that do not consume electricity while on standby. For electrical appliances that require timed startups, ultra-small sized batteries similar to those used in personal computers could be fashioned, and all unnecessary clock displays could be removed.

The expenditure for standby mode electricity consumption in households is shown in Table 3-7. It costs nearly 10,000 yen (U.S.$83) annually. The cumulative cost over a five-year period would be enough to pay the additional price for appliances which do not have standby electricity consumption. Consumers often have no choice but to buy and use these electrical appliances without having any notice of the costs associated with the standby features. If electrical appliances that do not consume standby electricity are developed and promoted, 10 to 15 percent of household electricity consumption could be eliminated.

4.11 Increase of Junk Mail

E-mail is very low cost and has rapidly grown as an alternative to postal mail and fax, and energy consumption of e-mail seems lower than its alternatives. A simple click of the mouse can send the same files to many people simultaneously. A result of this efficiency however, is seen in the flood of "junk mail" which is also received. A certain increase of junk mail would not add much energy consumption, but junk mail causes a substantial loss of time. Software that can eliminate unnecessary e-mail by recognizing senders will eventually be required as a counteractive measure. Often in society if something is convenient, it will eventually become too convenient. Such software would work automatically to detect the accumulated levels of junk mail, eliminating them when the levels became too high. It remains to be seen, however, whether our society can cleverly manage and utilize the new efficient communication tool of electronic mail.

4.12 Acceleration of Time

IT allows information to be processed quickly and transmitted at high speed, and people utilize these technologies to save time. One consequence of higher speeds and less time to complete tasks is accelerated consumption of energy and resources. The speed of economic activity accelerates with the use of fast technologies, and society is faced with the potential for the entire economy run at a higher speed, as can be seen in international currency trading.

If shopping requires less of a consumer's time, consumption will automatically be accelerated. As the time to acquire goods or services becomes less, GDP in a certain period of time will expand. It usually requires some time and effort for a consumer to acquire catalogs or travel to shops, after the initial decision to make a purchase. IT and electronic commerce will shorten the time required to make purchases, accelerate the speed of our use of time resources, and satisfy the desire to do more things in a certain amount of time. These factors are correlated to the growth of economy. In ten years we may be able to realize the advances that would otherwise occur over fifty years. Whether this capability is seen as positive or negative remains to be seen. It is certain, however, that this acceleration will increase the number of options a person can choose from to accomplish in a lifetime.

Time has a special significance relating to energy consumption. A behavioral science survey in the United States found that people think they use energy to save time. One respondent replied, "I know that I could cut energy use tremendously if I had lots of time to spare. But who does? We're always busy. The more energy I use, the less time it takes to get things done. I gain time through the energy I use."[6]

The survey shows that the words "I am busy" are considered a sign of social status. People are willing to buy time if it is possible. The average person will not live for more than 100 years and would like to accomplish many things over that lifetime. This personal desire for accomplishment has implications for the study of energy consumption. It is impossible to directly deter people from consuming energy to "buy" time. It may be possible, however, to implement measures to improve energy efficiency or to tax energy resources and raise the price of energy in order to suppress or limit energy consumption.

IT shares a similarity with energy: both are used to save time. This is the significance of IT in the New Economy.

If energy efficiency improvements and decentralized energy systems described in the previous chapter are not available soon enough, the rapidly expanding economy will continue to grow with a dependence on fossil fuels, possibly causing the eventual extinction of our societies, just as the prehistoric dinosaurs disappeared.

5. ENERGY SYSTEMS IN THE NEW ECONOMY

There is a question as to what the appropriate energy option for the New Economy will be, but the energy option has basically no relationship to the development of the New Economy.

Through IT, better energy efficiency may result in an overall reduction of energy demand over the long term. This is related to the development of decentralized energy systems.

The energy demand of IT is not large. Peak energy consumption for a personal computer with peripheral devices is 200 watts for a desktop computer with CRT monitor and 30 watts or less for a notebook style personal computer with liquid crystal display. A solar photovoltaic panel can supply sufficient electricity for such demand. Ongoing innovations show that the future energy demand of the central processing units and memory used in IT will decrease further. Clearly, decentralized energy systems that use renewable energy sources will be able to power the electronic equipment of the New Economy.

Decentralized energy systems are advantageous in that they are capable of responding to the huge amount of energy demand being generated simultaneously throughout many places in the world affected by the New Economy. Decentralized energy systems have similarities to the network structure of the Internet, in that there is no requirement for a central administration headquarters. The reason for the fast penetration of Internet is well known as the very nature of its decentralized structure.

As mentioned above, the New Economy has the potential to become more energy efficient, while lessening energy demand with a transition toward a service economy and electronics oriented IT. Given that the per unit energy demand of electronic devices in the New Economy is not large, we will see a tendency toward decentralized energy systems. The shift to the New Economy will accelerate this trend.

There is however, an alternate view. Expansion of the New Economy could create rapid economic growth, which will negate the benefits of energy efficiency improvements and promote more consumption of energy and resources. Electronic commerce could promote consumption globally, and even if energy efficiency is better, the overall energy consumption could increase, in a way, stealing time from the future.

6. CONCLUSION

The New Economy has the potential to improve energy efficiency. It will facilitate the societal shift from the current energy-intensive economy, which relies upon the heavy consumption of raw materials, to an information-intensive economy with service and software industries. It is possible that energy consumption per unit of GDP would decrease. This decrease has never really been experienced before, and is a desirable direction for the world to aim for.

This chapter discussed the prospects and problems of the New Economy. The main prospects include dramatic decreases in the consumption of paper and transportation and the efficient use of energy through the application of information technology in a variety of industries. As with any transition, other problems will emerge. Some such problems include standby electrical consumption of appliances, efficiency aspects of small package transportation. Another problem may be the increase of electronic junk email due to its convenience.

The calculations used here provide only examples, but they do indicate that the Economy may be able to reduce the amount of CO_2 emissions into the atmosphere. If further developments produce IT with more energy efficiency, the energy saving will be even greater.

The energy demand of IT devices is relatively small and thus decentralized energy systems are suitable for them. Accordingly, the transition toward the New

Economy should be complimented by the promotion of the decentralized energy systems.

The New Economy will accelerate the growth of the economy and promote economic "globalization." This will make it possible to realize in ten years the economic conditions that would have otherwise required fifty years. If the rate of economic growth is too fast the installation of decentralized energy systems may not keep up with the New Economy, and as a result the fossil fuel based energy systems may become entrenched.

REFERENCES

[1] Tsukio,Y. 1996. *Sasutenaburu shakai e no sujimichi* (The Road to Sustainable Society). Toyokeizai Shinposha.

[2] Tsuchiya, H. 1994. "Energy Analysis on Daily Activities." Energy for Sustainable Development, *The Journal of International Energy Initiative*, India, Vol. 1, No. 3.

[3] Tsuchiya, H. 1998. "Information and Energy saving." *Journal of Electric Engineers*, Japan. Vol. 18, No. 4.

[4] Nippon Telephone and Telegraph Co. 1993. *Zaimu joho (Dai 1 ki - 7 ki)* (Financial Information (1st to 7th Term).

[5] Nippon Telephone and Telegraph Co. 1993. *Shuyou toukei shiryoshu* (Compendium of key statistics).

[6] Erickson, R. J. "Household Energy Use in Sweden and Minnesota: Individual Behavior in Cultural Context" in Energy Efficiency: Perspective on Individual Behavior, W. Kempton (edit.), *American Council for Energy Efficient Economy*, Washington D.C. 1987.

Chapter 4

IMPACTS OF THE INFORMATION TECHNOLOGY REVOLUTION: ECONOMIC CHALLENGES OF NEW DEVELOPMENT PATTERNS

YASUHIRO MUROTA
Shonan Econometrics

1. INTRODUCTION

This paper is an analysis from the perspective of how the information technology (IT) revolution will affect economic and political systems. The IT revolution represents such a large groundswell in technological innovation that it is called the second industrial revolution; it will define development patterns over the next 50 to 100 years. In that context this paper will discuss in particular the implications of IT for global environmental problems and developing countries.

The IT revolution has only just begun, making it too early to assess its full significance. However, in view of the depth and magnitude of its impact on society as a whole, it is valuable to speculate now on what it means to society and the phenomena it will bring about, even though our speculation may be incomplete.

Section 2 examines what the IT revolution is, and Section 3 touches upon its impacts on economic and political systems. Section 4 explores the IT revolution's effects on global environmental problems and developing countries, and Section 5 summarizes briefly by imagining what future societies based on IT will be like.

2. WHAT IS THE IT REVOLUTION?

This paper uses the term "IT revolution" to mean "the widespread use which began in the 1990s of digital goods through the integration of personal computers and the Internet."[1]

K. Matsushita (ed.), Environment in the 21st Century and New Development Patterns, 87–113.

This paper shall explain the characteristics of digital goods and then discuss the developmental process of personal computers and the Internet, which brought about the popular use of digital goods. The paper will then explore the characteristics of the IT revolution.

2.1 Characteristics of Digital Goods

Digital goods are a subset of information, and the term refers to that which can be digitized. Examples would be music (digitized recordings on compact disks) and a company's financial data (digitized so it can be used by spreadsheet software and other applications). Needless to say, steel beams and agricultural produce do not fall into this category. These will be called non-digital goods (see Section 3.1).

It is perhaps the definition of von Neumann's theory of self-reproducing automata that shows the features of digital information.[2] While this is a definition of information, it actually defines none other than digital goods themselves. The definition is:

(a) Machine-readable
(b) In a form that can be readily duplicated.
(c) Duplicated information has the same form as the original without destruction.

Thus digital goods can be duplicated, and copies are exactly the same as the original. In addition, duplication can be done at practically no expense (or without labor).

Digital goods therefore have the following characteristics.[3]

(a) There is (nearly) zero cost involved in duplication and transmission, the latter meaning the process of passing goods from sender to receiver.
(b) Use is non-excludable, barring humanly imposed limitations such as copyright, encryption, and digital watermarking. Fundamentally, therefore, there is no concept of ownership.
(c) Production of digital goods involves creating something unknown, so there is no stable production relationship between input and output (there is great uncertainty, and ordinarily inputs cannot be gauged monetarily).
(d) Generally copyability is perceived as a disadvantage, but a trait of digital goods is that they can be copied, and improved through copying (see Section 2.5.3).

Personal computers are used to produce and process digital goods, and the Internet is used to transmit them, which makes computers and the Internet the media of the IT revolution. Let us therefore begin by examining their developmental processes.

2.2 Developmental Process of Personal Computers

2.2.1 The Development of Personal Computers

The development of personal computers can be divided into three periods (Table 4-1).[4]

Table 4-1. Personal Computer Development.

Year	Event
1970	Ritchie develops UNIX.
1971	Intel releases Intel 4004.
1972	Ritchie develops C Language.
1974	Kildall develops CP/M2.0.
1975	Altair 8800.
	Bill Gates and Paul Allen put 4KB BASIC on a computer.
1977	Apple II, Commodore PET, TRS80.
1978	Intel releases 8086. X86 becomes world standard.
1979	Wordstar and Visicalc.
1981	IBM PC.
1982	Mouse can be used with IBM PC.
	Lotus 123.
1983	PC DOS 2.0 becomes industry standard.
	Microsoft C, Turbopascal.
1984	Apple Macintosh.
	GUI on Mac OS.
1985	Windows 1.0.
1991	VB receives Byte Magazine's excellent product award.
1992	Windows 3.1.
1995	Windows 95.

Source: Byte, "20th Anniversary", Sep. 1995

(a) Beginning Period: 1970s
This is when personal computers appeared. Elements of this period include the birth of the microprocessor (1971), emergence of the concept of a personal computer using it (combining CPU, memory, and software; 1975), appearance of personal computers as commodities (simultaneously by Apple, Commodore, and Radio Shack; 1977), and the emergence of full-fledged applications (the spreadsheet VisiCalc, the word processor Wordstar; 1979).

(b) Establishment of De Facto Standards: 1980s
With the arrival of the 1980s personal computers achieved market recognition and established themselves as tools for working. A de facto standard came when the IBM PC appeared in 1981. It had Intel's 8088 CPU and Microsoft's PC DOS, and as the specifications were made public, other manufacturers started making computers to those specifications, which became an industry standard. Because the spreadsheet application Lotus 123 (1982) could be used on the PC, this accelerated the business use of personal computers.
Improved CPU performance and lower memory prices during these years began the search for new specifications that emerged in the 1990s. Examples are introduction of the mouse (1982), graphical user interface (GUI) (1984), and Windows 1.0 (1985). Apple's use of the mouse-GUI combination in the Macintosh already in 1984 is well known.

(c) Fusion with the Internet: 1990s
Fusion of computers and the Internet began in these years, with a symbolic event being the 1995 appearance of Windows 95, making it simple to connect to the Internet (browsers will be covered below).
With the improvement of personal computer performance, they are now said to match that of earlier supercomputers,[5] and they are far easier to operate than mainframe computers have been until now, which is of course why they are so widely used.

2.3 PCs Bring Computers to the People

Unlike mainframe computers, personal computers are (1) low in price, (2) their software is easy to use, and (3) there is no usage fee, making them available to everyone. By way of note, an estimated total of 84.3 million units of PCs were shipped worldwide in 1997.[6] Assuming the stock to be twice that number, there were 160 million PCs in operation around the world at that time. This number alone shows that PCs were already in widespread use.

The spread of computers is perhaps easy to understand by comparing with the industrial revolution. Compare mainframes to the 1712 invention of the steam engine by Thomas Newcomen, and personal computers to James Watt's 1769 improvements in design. The difference between the two designs was that Watt's steam engine was highly efficient and could be used anywhere because it was not bound to locations where there was fuel. This, it is said, brought the steam engine into wide use. If mainframes are characterized by giant corporate users, computer experts, and high price, then the characteristics of personal computers are ordinary users, amateurs, and low price. Mass sales of PCs brought down the costs of CPUs and memory. Further, because anyone could use them, their widespread uşe was aided by a positive cycle in which software was convenient and anyone was free to develop software, which resulted in good software emerging from some unexpected places.

If IBM is the king of mainframes, then Microsoft and Intel are the kings of PCs, a fundamental difference shown by the fact that when IBM set out to sell PCs, it had to use Intel's CPUs and Microsoft's operating systems. The cultures of mainframe computers and personal computers are entirely different.

2.4 Developmental Process of the Internet[7,8]

Table 4-2. Development of the Internet.

Year	Event
1962	Licklider announces Galactic Network concept.
1964	Kleinrock publishes book on packet switching.
1965	Computers achieve remote packet switching connection for the first time.
1966	L.G. Roberts announces his ARPANET concept.
1968	DARPA provides funding for ARPANET. Kahn designs it.
1972	Kahn gives public demonstration of ARPANET at ICCC.
	The term internetting is used for the first time.
1973	Kahn and Cerf develop TCP/IP.
	Metcalf invents Ethernet.
1982	4.2 BSD supports TCP/IP.
1983	ARPANET fully adopts TCP/IP.
1986	NSFNET begins operation.
1989	WWW used experimentally for the first time (T. Berners-Lee).
1990	NSFNET absorbs ARPANET.
1991	Full-blown commercialization of the Internet (CIX backbone).
1993	M. Andreessen releases Mosaic.
1994	Search engine Lycos begins operating.
	Netscape Navigator released.
	Mac OS 7.5 supports TCP/IP.
1995	NSFNET gives management of backbone to private sector.

Source: ● J. Parnell. "Programming is fun," Nikkei Byte, September 1997 (in Japanese)
● Nikkei BP. Nikkei Digital Encyclopedia, Nikkei BP, 1998 (in Japanese)

2.4.1 Birth of an Idea: 1960s

Already in the early 1960s Licklider at MIT was speaking of a Galactic Network in which everyone in the world would be connected, and could use it to obtain data and programs from any site.[9] This fits the present definition of the Internet by the Federal Network Council (FNC).[10]

2.4.2 Birth of Elemental Technologies and Start of Experimental Network: 1960s-1970s

Some of the elemental technologies that make up the Internet are packet switching and TCP/IP. The former was developed by Kleinrock in the 1960s and made it possible to transmit data at low cost and with high reliability, while the latter is a protocol, invented by Kahn and Cerf in 1973, that connects computers to one another over a network.

Efforts to use packet switching on a computer network started in 1969 at UCLA as ARPANET, and this was publicly announced in 1972 at the International Computer Communication Conference (ICCC). Kahn, who designed ARPANET, based the network on open architecture, meaning that local networks connected to ARPANET could have any design whatever, so ARPANET was conceived as a meta-network in which those local networks were mutually connected. TCP/IP was the protocol devised for that purpose.

2.4.3 Development as an Academic Network: Mid-1980s through first half of 1990s

During 1980s ARPANET started to find practical applications for defense-related research and development, and 1986 saw the initiation of NSFNET, which used packet switching and TCP/IP. This was a network for institutions of higher learning, intended to connect supercomputers. In 1990 it absorbed ARPANET.

2.4.4 Popularization: Mid-1990s and later

Two more technologies were needed to bring the Internet to the people. These were the World Wide Web (WWW) and Mosaic. The World Wide Web, a wide-area information retrieval system in which many computers are connected as a network, was devised by Berners-Lee in 1989, while Mosaic is a "browser" allowing anyone to use the World Wide Web that was developed by Andreessen in 1993.

Commercial use began with email service at the end of the 1980s and quickly expanded in the 1990s when browsers came into general use. At the end of 1997 there were reportedly over 100 million Internet users.[11] The figure of 160 million personal computers given above roughly matches this number.

2.5 Fusion of Personal Computers and the Internet

When personal computers and the Internet were combined in the mid-1990s, the IT revolution began in earnest. Specifically:

(a) CPU and memory performance was augmented, and with the appearance of Windows 95 (1995) connecting to the Internet became a simple matter. In addition, modems and other connection equipment became cheaper and faster (about 33.6 kbps).

(b) Improvements in Mosaic made it usable to everyone. Mosaic's developer Andreessen became one of the founders of Netscape (Netscape Navigator came out in 1994), and Microsoft made its Internet Explorer browser a standard component of Windows.

(c) Providers sprang up around the world, allowing anyone to enjoy Internet services (e-mail, shopping, searches) as long as they have a personal computer, connection software, a browser, and a modem (the search engine Lycos began in 1994).

(d) Owing to these developments, NSFNET transferred its backbone to the private sector and was retired in 1995.

2.6 Characteristics of the IT Revolution

Some characteristics of the IT revolution are: (1) rapid evolution, (2) instead of being a major project of governments or giant corporations, its main driving force is cumulative ingenuity of individuals, (3) economic motives are not the primary driving force, and (4) it can be described as creative destruction.

2.6.1 Rapid Evolution

The term "dog year" is sometimes used in connection with the IT revolution now in progress. One year in a dog's lifespan is equivalent to seven years for a human: the changes occurring in this field during one year would ordinarily take seven in other fields.

The pace of change is indeed fast. For example, the first microprocessor, Intel's 4004 (1971) had 2,300 transistors and its performance was 0.06 MIPS ("Most Important Chips.[12] Twenty-four years later in 1995 Intel's P6 had 5.5 million transistors and ran at 300 MIPS. In 2012 Intel's Micro 2012 is predicted to have 1 billion transistors and run at 100,000 MIPS.[13] This means that performance increased 5,000-fold in the first 24 years, and will increase 300 in the subsequent 17 years. This is as if a Cadillac that was selling originally for 5 million yen could now be purchased for 500 yen. The annual rate of change is 40%.

The Internet is likewise evolving rapidly. In 1993 there were under 5 million Americans using the Internet, but in 1997 the number was 62 million.[14] UUNET, a large U.S. backbone provider, says that Internet usage revenues double each 100 days.[15] Development of the Internet has accelerated especially since the mid-1990s when commercialization began.

In a mere three decades both personal computers and the Internet have had great impacts on society, and developments have proceeded faster than in the industrial revolution. With electricity, for instance, more than a century passed from the time of Faraday's experiments (1831) to the widespread use of electricity in factories and homes (80 diffusion in the U.S. in 1939).[16]

2.6.2 Individual Ingenuity

The key to advances in both personal computers and the Internet is individual ingenuity, as they did not begin as projects by a government or other body. In this sense they differ totally from, for example, NASA's missions to the moon.

Technological innovation ordinarily makes people think of giant corporations or specialized technical organizations as the promoting institution, technical experts as the people pursuing it, and the necessary sophisticated and expensive equipment, but the IT revolution is quite different because the agents are individuals, and new technologies have been pioneered by a combination of non-experts, garage workshops, hand-made devices, and the like. In this field large corporations like IBM had no choice but make use of others' ingenuity (as noted above, the IBM PC used Intel's CPU and Microsoft's OS, and sales increases depended on Lotus 123), and in only a few decades the companies founded by such individuals transformed into the giant corporations that now sustain the IT industry.[17]

Following are snapshots of some individuals who founded IT.

Bill Gates: With Paul Allen he wrote 4KB BASIC for the first personal computer, the Altair 8800, in 1975. Later he put MS-DOS on the IBM PC, made it the industry standard, and worked on fusing personal computers and the Internet with Windows. Born in 1955, he began working with a teletype at school at age 12. Dropped out of Harvard after two years and founded Microsoft. Currently worth $50 billion.[18]

Steve Wozniak: Designer of the Apple II. In 1975 began making computers in his father's garage with Steve Jobs. A consummate hacker. Now teaches computers to elementary school children.[19]

Tim Berners-Lee: Creator of the WWW. A British-born physicist, in 1989 he applied to his employer (European Laboratory for Particle Physics, CERN) to get $80,000 for WWW development. Despite the huge amount of information in computers, he felt that it was not well organized, and therefore intended to develop a simple protocol that would conduct network searches and provide links. He is presently the director of the WWW Consortium.[20]

Marc Andreessen: Began programming at age 12 with a Radio Shack computer. While a student at the University of Illinois he was hired as a programmer for $6 an hour. Developed Mosaic with Bina there. In 1994 he founded Netscape Communications with James Clark.[21]

As these profiles indicate, it was the vision and creativity of individuals that developed the key technologies (even the contribution of Berners-Lee was not related to his job). These technologies then attracted the necessary funds, and the developers nurtured their companies. The key technologies were not developed by experts with sophisticated, specialized knowledge using vast expenditures by major corporations or research institutes. Governments played a passive role in these advances by not becoming involved.

Interesting, however, is the existence of a research institute that was a source of knowledge needed in advancing the IT revolution, and that was the Palo Alto Research Center (PARC, <www.parc.xerox.com>), which developed the GUI, Ethernet, object-oriented programming (SmallTalk), the client server system concept, and the like. PARC was founded in 1970 for the purpose of "creating the architecture of information."

The unusual thing about this institute is that nearly all the ideas developed there were commercialized elsewhere. For example Steve Jobs, the creator of Apple's Macintosh, says he got the ideas for GUI and object-oriented programming here.[22] And the inventor of Ethernet later founded ThreeCom Corp.

2.6.3 Economic Motives Not Necessarily the Driving Force

While on the one hand the IT revolution made people like Bill Gates super-rich, one prominent idea is that excellent software is created none other than outside economic mechanisms. Indeed, a number of prominent software items have been developed in line with this thinking. Dan Bricklin, the inventor of spreadsheet software (who developed VisiCalc in 1979) did not apply for a patent because he hoped that the lack of a patent would help advance this software field,[23] and indeed one could argue that this elicited the birth of superior spreadsheet applications like Lotus 123 and Excel (if not, he could have monopolized this idea for 17 years). And WWW inventor Berners-Lee says that he wanted browsers to be something everyone could use.[24]

It is the GNU project that made this thinking into a philosophy and put it into practice. GNU was started in 1984 by MIT's Richard Stallman for the purpose of providing a "free" UNIX-based operating system. LINUX is a well-known component.[25] Characteristics are:

(a) "Free" has more the sense of "unrestricted" than "no cost."
(b) Free software has the following requirements.

- Program content can be studied (i.e., it includes the source code), and can be modified to suit one's own needs.
- You may give a copy to your neighbor.
- You can freely modify the program.

In other words, these are the conditions for developing truly good software.

(c) "Free" here does not mean without cost, but is a counter-concept to "price." That is to say, Stallman contends that digital information technologies are not compatible with economic mechanisms (price and ownership) because copies can be made as one wishes and modifications can be easily made. As we saw with the definition of digital goods, this point cuts to the heart of the issue.

(d) Stallman came up with the concept of "copyleft" in order to widely disseminate such software. A concept that corresponds to copyright, copyleft was conceived to make software "free" (i.e., someone might modify shareware and sell it, so copyleft prevents this). Software may be used by registering it and adhering to the above conditions on free use.

Copyleft creator Stallman had joined MIT in 1971 and worked with a group that used only freeware, but in the 1980s he began this project because all software became proprietary. This work is now carried on by the Free Software Foundation (FSF), which sells a GNU CD-ROM and receives contributions from Intel, Motorola, TI, and others.

Similar thinking is at work in the Apache HTTP Server Project. Apache is a Web server made by improving a program written by the National Center for Supercomputing Applications (NCSA, which also created Mosaic) at the University of Illinois, and may be used for free.[26] Apparently 40 to 50% of WWW servers throughout the world now use Apache.[27]

2.6.4 Creative Destruction

The IT revolution is innovative, i.e., it involves creative destruction. This term was coined by Joseph Schumpeter, who explained it in the following way: "The position of the ideal state of equilibrium in the economic system, never attained, continually "striven after", changes, because the data change. No matter how many mail coaches are continuously added, they can never match the railroad."[28] In somewhat simpler terms, general purpose computers differ from personal computers.

What is more, new technologies have different proponents, which is Schumpeter's definition of the entrepreneur: "in general it is not the owner of stage-coaches who builds railways."[29] Indeed this is just as we saw above. It is a new group that brought about the IT revolution.

3. SOCIAL IMPACTS OF THE IT REVOLUTION

This paper now shall focus on the economic and political impacts of the IT revolution.

3.1 Impacts on Economic Activities

Goods are divided into two categories as follows.[30]
(a) Digital goods. New knowledge (music, scientific knowledge, etc.), market information (money: exchange function, scale of value, contracts, etc.[31]
(b) Non-digital goods. Motor vehicles, steel, agricultural produce, etc.
Thus the IT revolution means a far more rapid expansion of digital than non-digital goods in economic activities. Here this effect shall be referred to as the "digitalization" of the economy.

3.1.1 Consequences of Digitalization

As Schumpeter would put them, these consequences are:
(a) The birth of new ways for production and sales: the existing economic system becomes more efficient, and the economy improves.
(b) Departure from economic mechanisms.

These two have opposing effects on the market economy, i.e., while the first makes the market economy more efficient, the second promotes the dismantling of the market. These are described below.

(a) Heightened Efficiency and Vitality of Existing Economic System

New ways of producing and selling are possible through electronic commerce (EC), which may be thought of as commercial transactions through an electronic network (the Internet). Three characteristics of EC are new modes of production, new marketing methods, and economic vitalization. These are described below.

New modes of production: Global Division of Labor

With production, for example, a new mode of production is coming into being. For instance, U.S. automakers have succeeded in shorting the time span of vehicle development by network linkage of worldwide development centers with part suppliers.

In the early 1980s it took the U.S. auto industry an average of four to six years to develop a new vehicle. By contrast, Japanese automakers managed it in three years through the use of *keiretsu* (companies in the same corporate family) and other means. Adoption of electronic commerce by U.S. automakers has shortened that period to 30 months, and they are now trying to further shorten that to 24 months.[32]

In addition, in personal computer production built-to-order (BTO, i.e., producing after orders are received) is now a matter of course (as opposed to build-to-stock, or BTS). This became possible owing to network (Internet) linkage of buyers, factories, part suppliers, and distributors. Advantages are: (1) direct sales eliminate commissions for sales outlets, (2) no need for excess part inventories, (3) buyers can obtain products with the desired specifications, and (4) shorter time from order placement to product delivery. PC makers that adopted this system are increasing their market shares,[33] and other makers (Compaq, for example) are rushing to adopt it.

New Marketing Methods: Virtual Stores

New ways of selling goods are possible through "virtual stores," in which there is no physical store (the bookseller Amazon.com, etc.), general insurance sales (InsureMarket by Quicken[34]), and the like. Since the former has no physical store, there are none of the associated costs (acquisition of retail space, personnel, product inventory). From the buyers' point of view, virtual stores offer a far greater selection than conventional stores (Amazon.com has 2.5 million available titles, while conventional stores have at most 250,000[35]), and although shipping costs are involved, the needed books can be delivered in a short time. What is more, Amazon.com carries both new and used books, allowing buyers to obtain all their needed books there.

InsureMarket is a subsidiary of Intuit, which develops accounting applications for individual users. Here the consumer can quickly obtain estimates from multiple insurance companies for a variety of insurance types (including life, auto, home, and income). For example, for term life insurance InsureMarket offers estimates from a maximum of seven companies, as well as third-party ratings (Moody's, Standard & Poor's, etc.) for each insurance company (owing to regulations, people must buy policies through their local insurance agencies).

Economic Vitalization

These changes are creating good business conditions, especially in the U.S. where they started. For example, a Department of Commerce report summarizes IT impact on the U.S. economy as: (1) supporting 25% of U.S. economic growth in recent years, and (2) lowering inflation by 1% in recent years.[36,37]

(b) Departure from Economic Mechanisms

It is a fact that the appearance of these new products, i.e., personal computers, and the increased efficiency of existing information transmission mechanisms by the Internet have expanded the economy and made it more efficient, but we must not disregard another consequence: The IT revolution is oriented toward de-marketization, which is because of the very nature of digital goods. Here we consider two specific reasons.

Production does not necessarily have an economic motive.

The GNU project's Stallman says that the conditions necessary for good software (free software) are that it can be decoded, people may give copies to their neighbors, and people may freely make improvements. He makes the following statement: "Digital information technology contributes to the world by making it easier to copy and modify information. Computers promise to make this easier for all of us."[38]

This is only natural assuming the definition of digital goods in Section 2.1. Digital goods, in other words, are basically non-market goods. Stallman mimicked the Manifesto of the Communist Party with the "GNU Manifesto" that explains this point as follows:

(i) Because I like a program I want to share it with other people who like it.
(ii) Everyone is permitted to modify and redistribute GNU (it is kept out of the public domain in order to prevent someone from using it to make commercial software, which could not be redistributed, i.e., copyleft).
(iii) Many programmers are happy about GNU and want to help.
(iv) Many programmers are unhappy about software commercialization because they cannot regard other programmers as comrades.
(v) To programmers, copying programs is as natural as breathing.
(vi) Many people write programs without a monetary incentive. Good programmers feel an "irresistible fascination" for writing programs.

The Apache Project says more or less the same thing.[39]

(i) Apache is robust and has a commercial-grade HTTP protocol. Individuals and institutions can build reliable systems on it (in other words, it's not a toy just because it's free).
(ii) Because WWW protocols are un-owned, the Web is a level playing field.
(iii) Apache is an "organic entity," so the people who benefit often contribute to it by making improvements, fixing bugs, and offering support for others.

Efficiency in Resource Allocation Is Not Dependent on the Market

As mentioned along with the definition, as digital goods have a strong character as "public goods" and of "externality," distribution by the market does not assure efficiency.[40]

And although the market mechanism concentrates information in price,[41] this no longer means very much under the IT revolution. It will no longer be necessary to concentrate on the single type of information called price. The market mechanism was an innovative way of effectively allocating resources when conveying information required too much money and time. We could say that in a way one-to-one marketing basically negates the pricing mechanism.

3.1.2 The New Economy

In the U.S. there is already a controversy over the economic impacts of the IT revolution. Specifically, does the IT revolution have a large impact on economic activities? In simple terms, new economic theory says that owing to the IT revolution the U.S. has begun a new age of prosperity. Arguments agreeing and disagreeing with this view are as follows.

Agree:[42]

(a) In recent years the U.S. economy has enjoyed high growth, low unemployment, and low inflation, a phenomenon that has never before happened.

(b) This is because U.S. companies stole a march on the rest of the world by adopting advanced technologies (in the form of capital investments). In this way the U.S. is beginning investment-driven economic growth, and it will probably continue.

(c) The reason for this advance is that U.S. companies were structured in a way making it easy to incorporate advanced technologies. Specifically, their flexible structures allowed them to adapt quickly to changes, and the financial system had been reformed in a way that was amenable to fostering such industries.

Disagree:[43]

(a) It is a fact that the U.S. economy was good in 1997, but this was nothing outside ordinary economic fluctuations. That is to say, it was merely a phenomenon that can be easily explained with Okun's law. Considering that only 10 years ago people were claiming that the U.S. economy was in ruins, new economy theory is a mere superficial view.

(b) It is true there is no inflation in spite of declining unemployment, but this is because there is lower pressure for rising wages.

Ordinarily a positive analysis would settle such a controversy, but at least a look at current statistics does not show clearly what effects the IT revolution has on macroeconomic productivity. This has brought the controversy to a stalemate.

In light of our analysis here, however, both these arguments seem one-sided for two reasons.

(a) Considering the new projects that have emerged and the innovation in distribution that has occurred due to networking in the IT revolution, there is no doubt about the large positive economic effect.

(b) But fundamentally the IT revolution has a "creative destruction" aspect, and there is an especially great possibility that it will emerge in the form of de-marketization (the superiority of digital goods over non-digital goods). Thus it will destroy existing organizations and decrease sales of existing products. If these aspects are taken into consideration, it does not make much sense to measure the effect with GDP, an aggregate value.

3.1.3 Implications of the Asian Crisis

The Asian crisis is perhaps a good example of balance having been lost between digital goods (for instance, capital for speculation), for which speed can be adjusted quickly, and non-digital goods (such as merchandise imports and exports), for which speed is adjusted slowly.[44] It was U.S. Federal Reserve Chairman Alan Greenspan who incisively observed that the Asian crisis was a result of high-tech IT advances.[45] His points were as follows.

(a) Recent technological innovation (i.e., the IT revolution) has achieved substantial innovation for financial systems as well, creating a "new high-tech international financial system."

(b) This has resulted in substantially lower costs for capital transactions across national borders, making for brisker international capital movement.

(c) On the other hand, it has imposed quick and decisive penalties on countries with economic weaknesses. It is also very easy for financial disturbances to spread.

(d) The first indication of that was the Mexican crisis (1994-95), and the second was the Asian crisis (1997-98).

That is to say, the IT revolution's advantage is improved economic efficiency by globalization, but owing to the characteristics of digital goods (instantaneous transferal), the revolution brings instability to the existing economic system. These will continue to appear in various places as transitional phenomena.

3.2 Impacts on Political Systems

Key words for the discussion of political systems are "e-democracy" and "e-government,"[46] which are becoming reality owing to the IT revolution. Specifically they signify the revival of direct democracy and the shift to electronic government administration.

3.2.1 Revival of Direct Democracy

Achievements of the IT revolution are also changing political processes. With increases in the number of people participating in political discourse and resolutions, democratic states in the past shifted from direct to indirect democratic systems. The people voted in elections to choose parliamentary representatives, who debated and decided on policy in parliament.

It is possible that the IT revolution will upset this trend. Once again all people will have the opportunity to directly make their opinions known and participate in decision-making. This is already appearing in its early stages.

Example 1: Telephone Referendum

In April 1998 the city of Thousand Oaks, California held a telephone vote for the citizens over land development plans.[47] Each of four plans had been assigned its own telephone number, so citizens could vote for the one they preferred by dialing its number, and express their opposition by not dialing it. Citizens who were displeased with the whole menu could call a fifth telephone number and leave a message. Set up with the cooperation of GTE, the system would not allow multiple votes to be cast from the same number.

Local community antenna television (CATV) served as a supplementary information source by showing a 27-minute video describing each of the project plans in more detail. In the next stage Thousand Oaks apparently plans to use the Internet as a means for citizens to discuss what they want.

Example 2: Virtual Public Hearing

In February 1998 the U.S. Senate Small Business Committee (chaired by Senator K. Bond) held a "virtual" public hearing on the reform of the Internal Revenue Service.[48]

This hearing combined teleconferencing and the Internet. First of all, the three cities of Washington, St. Louis, and Salt Lake City were linked by teleconferencing, allowing people in each place to testify without having to go to Washington. Second, the hearing's progress was reported over the Internet (www.senate.gov/sbc), allowing the citizens to both see it and use the chat room to interpose their own opinions.

Example 3: Discussion of Budget Plan in Brent Council (England)

Brent is a city on the outskirts of London with a population of 240,000. In January 1997 an experiment was organized by UK Citizens Online Democracy (UKCOD) in which the city's residents participated via the Internet in deciding on the proposed budget.[49] This involved first showing the city's proposal and then comparing figures from the same categories for nearby cities. Finally there were answers to questions about appropriate tax levels and spending allocations, which were sent by e-mail. This gave the citizens far more latitude and options for expressing their opinions than a simple vote.

Initiatives like these are appearing around the world, with the most famous of all being that of Minnesota E-Democracy in the U.S. state of Minnesota.[50] Of course these experiments have only begun, but there is a greater possibility that, by using IT, the need to depend on representative democracy will disappear. Needless to say, it will be necessary to provide options and to have ways to tally and categorize opinions. Nevertheless, this represents the passing of the time when people use elections to express their will once in several years. In other words, the time may come when we will no longer have need of professional politicians, a function that was a product of representative democracy. Furthermore, there will no longer be much necessity for participants to stay within the confines of a single region or country. That is to say, voting rights currently belong to adults of the present generation who live in a certain area, but when voting results affect later generations or other geographical areas, those people have not been given voting rights. The IT revolution could change this situation too. Globalization is affecting politics.

3.2.2 Electronic Administration

It is said that governmental administration always grows. Government budgets and the number of public servants continue to increase each time a new problem emerges, such as public welfare or unemployment. Perhaps this is because existing mechanisms do not trim down agencies that are no longer needed. Making administration electronic might reverse this trend.

Example 1: Malaysia's Multimedia Super Corridor Scheme

One primary element of this scheme is e-government.[51] Ideas for electronic conversion of government services include:

(a) Digitizing documents
(b) Electronic procurement
(c) Linking all public agencies with networks

Example 2: E-Government in Britain

Britain is considering the following types of e-government.[52]

(a) Tax return filing, passport application, and the like performed by email.
(b) Such things are already being tried in Natwest, where self-managed businesses are registered electronically. Formerly the self-employed had to fill out four forms and file them with four offices.
(c) The British government is building a government intranet that within three years is to connect 500,000 civil servants.

4. THE IT REVOLUTION AND NEW "DEVELOPMENT"

4.1 Impacts of the IT Revolution

The IT revolution has the following three social impacts.

(a) Makes activities non-economic: de-marketization
(b) Moves from representatives and grouping to de-mediation and individualization (e-trade, e-democracy)
(c) Global integration: loss of nation-state framework

De-marketization is an inescapable feature of digital goods (the uncertainty that is a concomitant of production, the non-exclusiveness of consumption). Further, individualization can be expressed in ways such as one-to-one marketing and direct democracy. That is to say, media such as prices (the market) and professional politicians (parliaments) will no longer be needed. Networks have transcended the nation-state framework. Let us now explore what impacts these things will have on global environmental problems and developing country issues, which are crucial factors in considering "development."

4.2 The IT Revolution and Global Environmental Problems

Here the discussion shall narrow to global warming because it is representative of global environmental problems and is a critical one.

The issue of global warming could be defined as follows (for example, see Cooper (1998))[53]: "A long-term warming trend of the Earth begins due to emissions of anthropogenic greenhouse gases (GHGs). This brings about climate change, which could jeopardize the continued existence of humanity." Solving this problem involves the following difficulties.

(a) Consequences are uncertain (we cannot be sure when and how it will occur, or how it will affect human society).

(b) Those who pay the costs and those who benefit might be different. As the IPCC points out,[54] keeping GHG concentrations at their present levels would require, for example, reducing CO_2 emissions to about 60% of present emissions. But considering the future economic development of developing countries, even maintaining the present level of emissions would be impossible unless the emissions of developed countries were reduced to zero in the latter half of the twenty-first century.[55] Thus, there is a great possibility of a conflict of interests over this problem between developed and developing countries, or between present and future generations.

(c) It is a worldwide problem, i.e., global warming affects the whole world, and its effects vary according to region.

How will the IT revolution affect this problem?

4.2.1 Effects of De-Economization

The IT revolution could surprisingly turn out to be a way to solve global environmental problems. This is very much like how the contradiction of capitalism (pauperization of the workers) was "solved" not by the collapse of capitalism (as Marx predicted), but by economic growth. Emissions of CO_2, one cause of global warming, are more or less proportional to fossil fuel consumption, but it is possible that as IT transforms the economy, energy consumption will fall substantially, which would in turn cut GHG emissions.

Figure 4-1. World CO2 emission.

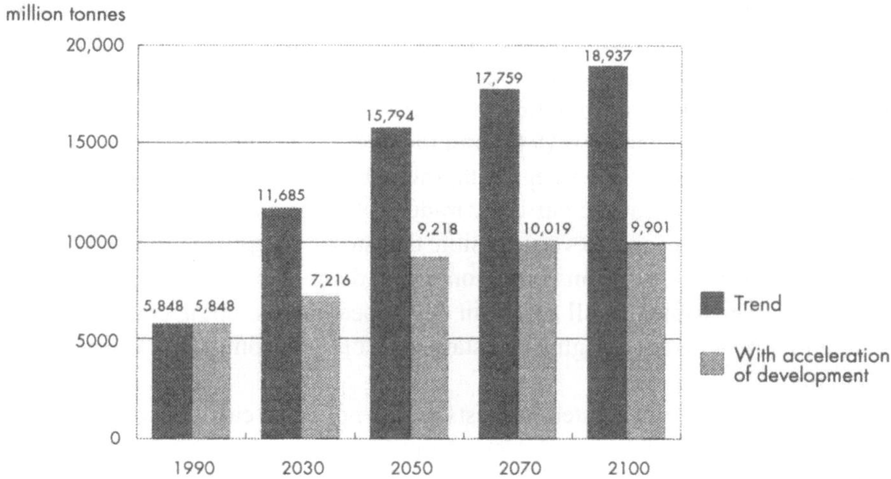

Specifically, this process might happen in this way.

(a) De-industrialization: This would cause a large reduction in the proportion of energy-intensive industries (steel, chemicals, cement, etc.), thereby greatly reducing the consumption of industrial energy. For example, the energy intensity of steel is 37.9 (Japan, 1995, kcal/yen), but it is only 1.8 for the machinery industry (calculated using statistics from The Energy Data and Modeling Center[56] and the Economic Planning Agency[57]). Accordingly, if these industries produce less, it will lead to reduced energy consumption.

(b) Communications replace transportation: A considerable amount of transport will be replaced by information. For instance, there might be much less need to commute to work. In the U.S. 7 million people now work at home in "virtual" offices, and it is predicted that the number will swell to 15 million in a decade.[58]

(c) Increasing electronic commerce: For example, banks are beginning to attract customers to online banking because it is cheaper than doing business in person at banks (Florida's Barnett Bank gives a discount for online customers, and First Chicago NBD now charges for window service.[59] This also allows customers to comprehensively manage their assets (integration of check payments, investments, etc.; Bank of America website, etc.[60]). Banks will no longer need branches, thus reducing energy consumption for operations.

4.2.2 Effects of De-mediation/Individualization and Global Integration

It is possible that an organization transcending national borders (a world government?) that will deal with global problems because this problem transcends the interests of one country or the present generation. In this situation we find similarities to the postwar founding of organizations for international cooperation such as GATT (established in 1947) and the IMF (established in 1945) to correct the negative impacts of earlier beggar-thy-neighbor policies (rise of totalitarianism, the global depression caused partly by trade protectionism).

In this case, however, it is very possible that such an organization would provide a venue for political decisions based on direct democracy, i.e., the IT revolution could make it possible that all adults, in developed or developing countries, would be able to vote with equal rights and state their opinions on issues. Consequences would be:

(a) Developed countries' vested interests would not be so easily recognized.
(b) Everyone would obtain the latest information, making it difficult to maintain positions based on biased information.
(c) International political parties would emerge.

4.3 The IT Revolution and Developing Country Problems

The developing country problem refers to a situation in which affluent developed countries and poor developing (underdeveloped) countries both exist in the world.[61]

According to Nishikawa[62] there are two ways to view "underdevelopment".

(a) Underdevelopment as belated development: In this view, developed countries have advanced development, while the underdeveloped countries have delayed development (e.g., Rostow's theory of development stages).
(b) Underdevelopment as a result of development: In this view, development of the developed countries induced the underdevelopment of the developing countries (theory of dependence).

4.3.1 Effects of De-Economization

Possibility of leapfrogging

Owing to the IT revolution information reaches all parts of the world equally and instantaneously. Taking advantage of IT revolution achievements quickly will make it possible for developing countries to build IT social structures instead of going through the steps followed by developed countries, i.e., from agriculture to industry to services (information). China, for instance, is clearly aiming for this. An expert at the Shanghai Academy of Social Sciences made the following points.[63]

(a) The IT revolution will bring about far-reaching changes in the industrial structure.

(b) China should abandon the industrialization strategy followed until now, and adopt a development strategy built around the knowledge economy.
(c) With the establishment of an information highway, underdeveloped regions of China will skip industrialization and immediately enter the information age. An example is Hainan Island.

If this is successful, it is possible that developing countries' energy consumption will decline because they will skip industrialization, thereby reducing their burden on the global environment. A simulation by Murota and Itoh using a world model showed that if developing countries succeed in accelerating their development, global CO_2 emissions in 2100 will be halved from 18.9 billion tons (carbon equivalent) to 9.9 billion tons.[64]

4.3.2 Effects of De-mediation/Individualization and Global Integration

The developing country problem as it has existed until now may very well disappear.
(a) National borders will lose their significance. A major premise of international trade is that labor does not move, but this means nothing due to the formation of networks because people can work anywhere (for example, U.S. companies use India as a center for software development). A U.S. Commerce Department report made this statement from the standpoint of developed countries (who should protect their interests): "Countries that have an insufficient supply of skilled workers will see high-skilled, high-paying jobs migrate to countries that can supply the needed talent."[65]
(b) One basis of dependency theory assumes the worsening of trade terms when developed countries export industrial products and developing countries export primary products.[66] But the IT revolution will eliminate these very labor division systems because there is nothing unusual about cutting-edge software being created in the rural areas of a developing country.
(c) And if a world government is formed and everyone has an equal voting right, it is more likely that developing countries' arguments will prevail.

5. FROM INDUSTRIAL TO INFORMATION TECHNOLOGY SOCIETIES

5.1 Characteristics of Industrial Society

(a) Economic growth is sustained by the heavy and chemical industries, and by the production of durable goods like automobiles.
(b) Developing countries maintain their economic activities by offering primary products, and by exporting components and labor-intensive products using cheap

labor, but transaction conditions put them at a disadvantage against developed countries.

(c) Economic growth based on industrialization increases energy demand, which causes global warming.

5.2 Characteristics of IT Society

(a) The concept of economic growth (GDP) no longer has much currency, with the reasons being that there is no longer any meaning in aggregates added together with price as weight because principal activities are outside the scope of the market, and that country-by-country aggregates no longer mean anything because the nation-state framework no longer applies. Accordingly, the practice of measuring the health of the economy with economic growth rate will be abandoned. In Japan it was only in 1955 -- after the rapid-growth phase started -- when national income statistics were kept and economic growth began attracting everyone's attention.

(b) The concept of international trade will also lose its meaning. The underlying premise of international trade is that labor is an immobile production element, but a world network eliminates this restriction (i.e., no matter where people live, they can work for a company in any country). And because the nation-state concept will become indistinct, the concept of commercial transactions across national borders will disappear.

(c) Compensation for labor has heretofore been paid for the restrictions imposed by time and place, but those restrictions will no longer apply (work may be performed anywhere, anytime). Further, the sole measure of work will be the end product.[67]

(d) IT changes society rapidly. In view of its characteristics, it proceeds with particular rapidity in the economic sector. In the political sector, however, institutional obstacles (the nation-state framework, sovereignty, etc.) require time for IT to work its changes. But the Soviet/Eastern Europe collapse shows it is possible that a political framework might seem permanent up to a certain point, and then change in an instant.

(e) Deregulation[68]: Normally one reason that governments impose regulations is to prevent the unequal distribution of information, but since progress is fast in the IT revolution, governments generally do not have the needed information. In that sense, for the "mal-distribution of information" we have recourse to third parties (for example Moody's and Standard & Poor's,[69] or for motor vehicles Autosite, Edmund's, etc.[70]). Also, consumers will use the wide variety of information made publicly available on networks. Industries subject to regulation have dull receptivity to technical innovation and are slow to adapt, which could be fatal as the IT revolution proceeds. Because of this, in many cases the concept of "natural monopoly" will no longer hold up under world integration and de-economization.

5.3 Problems of the IT Society

It is of course possible that the IT society will involve a variety of problems.

(a) Some people cannot keep up with change: the elderly, or people in developing countries who do not have sufficient information.

(b) Privacy violations: IT use could destroy the privacy of people using it.

(c) Concentration of information in certain groups: dispersion and concentration are two sides of the same coin. Information control may concentrate into the hands of groups in unexpected ways.

(d) Problems arising due to differences in speed of coordination between the IT sector and other sectors: the Asian financial crisis is one example of a gap, which emerged.

NOTES AND REFERENCES

1 This definition resembles that in the recently released *The Emerging Digital Economy* by the U.S. Department of Commerce (Apr. 1999; p. 8) in that both see the IT revolution as a fusion of PCs and the Internet. However, whereas the Commerce Department definition emphasizes the Internet, this report sees both elements as mutually complementary parts, and emphasizes popular use.

2 Noguchi, Yukio, *Joho no keizai riron* (Economic theory of information), Toyo Keizai, 1974 (in Japanese).

3 based on Chapter 3 of above.

4 based on *Byte*, "20th Anniversary," September 1995. (http://www.byte.com)

5 The most advanced personal computer at the time of writing (a Compaq with a dual Pentium) was reportedly more powerful than a 1985 Cray X-MP (J. Parnell. "Programming is Fun," *Nikkei Byte*, September 1997 (in Japanese)).

6 Nikkei BP. Nikkei Digital Encyclopedia, Nikkei BP, 1998 (in Japanese).

7 ISOC. "A Brief History of the Internet," Ver. 3.1, February 1998. (http://www.isoc.org)

8 Cerf Vinton, "The Internet Phenomenon," January 1998 ver. (http://www.cise.nsf.gov)

9 ISOC. "A Brief History of the Internet," Ver. 3.1, February 1998. p.2

10 The definition is as follows (according to ISOC. "A Brief History of the Internet," Ver. 3.1, February 1998, by the Federal Network Council, 1995; p. 15): The Internet is a worldwide information system that (1) has unique addresses under IP (Internet Protocol or a successor) that are connected with each other logically, (2) in which communication is supported by TCP/IP (Transmission Control Protocol/Internet Protocol or a successor), and (3) in which all users, whether private or public, can use communications and sophisticated services built on related infrastructure.

[11] U.S. Department of Commerce. The Emerging Digital Economy, April 1998. p.8

[12] *Byte*, "20th Anniversary," September 1995. (http://www.byte.com)

[13] Moore, G. "The Continuing Silicon Technology Revolution," Intel Archives. (http://developer.intel.com/solutions/archive/issue2/feature.htm)

[14] ISOC. "A Brief History of the Internet," Ver. 3.1, February 1998. p.2

[15] ISOC. ibid.

[16] David, Paul. "The Dynamo and the Computer: A Historical Perspective on the Modern Productivity Paradox," *AER*, vol. 80, no. 2, 1990, pp. 355-361.

[17] For example, while IBM's 1997 sales were $78.5 billion, Microsoft had $11.4 billion, Intel had $25.1 billion, and ThreeCom (founded in 1979 by Ethernet inventor Metcalfe) had $5.4 billion (according to WSJ Interactive Edition). New companies with at most 20-year histories have already grown to where their sales can be compared to those of IBM.

[18] Linton Weeks, "Money Magnates: At Opposite Ends of a Century, J.D. Rockefeller and Bill Gates Converge," *Washington Post*, June 18, 1998.

[19] *Byte*, "20th Anniversary," September 1995.

[20] Paul Andrews, "Web's Creator Still Gives It Momentum," *Chicago Tribune*, July 6, 1998.

[21] Steve Hamm, "The Education of Marc Andreessen," *Business Week*, April 13, 1998.

[22] *Byte*, "20th Anniversary," September 1995.

[23] *Byte*, ibid.

[24] Paul Andrews, ibid.

[25] GNU Project, "GNU's Not UNIX!" (www.gnu.ai.mit.edu)

[26] Apache Project, "About the Apache HTTP Server Project." (http://www.apache.org)

[27] *Nikkei Sangyo Shimbun*, July 22, 1998, p. 2.

[28] Schumpeter, J.A. The theory of economic development, Transaction Publishers, 1983. pp.61-62

[29] Schumpeter, J.A. ibid. p.67

[30] Goods are things that give a sense of utility when people use them. When limited with respect to unlimited desires they are economic goods, and when not they are free goods.

[31] Noguchi, Yukio, ibid.,

[32] U.S. Department of Commerce. ibid. p.17

[33] U.S. Department of Commerce. ibid. Appendix 3, p.5

[34] U.S. Department of Commerce. ibid. Appendix 4, p. 47

[35] U.S. Department of Commerce. ibid. Appendix 5, p.9

[36] U.S. Department of Commerce. ibid. pp. 5-66

[37] This conclusion is a little too optimistic because measurements are performed in a way that makes impacts look bigger. For example, regarding economic growth it compares the growth rate obtained by subtracting the IT-related

component from GDP with that including IT (i.e., actual GDP). But if there were no IT, people would have made up for that in other ways, such as by using adding machines. Although adding machines would contribute less to growth than IT, they would conceivably contribute more to growth in some sectors than in others (such as agriculture).

38 Stallman, R. "Why software should not have owners," GNU project, "GNU's not Unix!" (http://www.gnu.ai.mit.edu)

39 "Why Apache is Free," Apache Project, "about the Apache HTTP server project".

40 Okuno, M and Suzumura, K., *Microeconomics* 1, Iwanami, 1994 (in Japanese). p.23

41 Okuno, M. and Suzumura, K., ibid., explain it thus: "Price sums up all the necessary information and quickly communicates it to all economic entities as public information. Thus, this is highly efficient as a mechanism for gathering and conveying information."

42 Zuckerman, A.B. "A Second American Century," *Foreign Affairs*, May/June 1998.

43 Krugman, P. "America the Boastful," *Foreign Affairs*, May/June 1998.

44 On the course of events in the Asian crisis, see IMF, *World Economic Outlook*, May 1998, p. 3 and Chapter 4.

45 FRB. "Testimony of Chairman Alan Greenspan," January 30, 1998. (www.bog.fed.us)

46 Also known as electronic town hall, this term is based on a slogan used by Ross Perot in his 1992 presidential campaign.

47 Engardio, J.P. "Direct-Dial Voting Options Too Vague, Some Residents Say," *Los Angeles Times*, March 26, 1998.

48 Mathis, D. "Weekly Congressional Column for Springfield," *Garnett News Service*, February 10, 1998.

49 UKCOD, " Brent Council - Budget Consultation." (www.democracy.org.uk)

50 Clift, S. "Democracy Is Online," *On the Internet Magazine*, March/April 1998.

51 Ismail Hashim, "Public's Role in Realising the E-Government Vision," *The New Straits Times*, March 23, 1998.

52 Philip Johnston, "Stop the Technology Madness: You Can Net some of the People Some of the Time," *The Daily Telegraph London*, June 2, 1998.

53 Richard Cooper, "Toward a Real Global Warming Treaty," *Foreign Affairs*, March/April 1998.

54 IPCC. Scientific Assessment of Climate Change, June 1990.

55 Manne, A.S. and Richels, G. "Global CO_2 Emission Reductions: The Impacts of Rising Energy Costs," *Energy Journal*, vol. 12, no. 1.

56 The Energy Data and Modeling Center. *Enerugii keizai toukei youran* (Handbook of Energy and Economics Statistics), 1997 Edition (in Japanese).

57 Economic Planning Agency. *Kokumin keizai keisan nenpou* (Annual Report on National Account), 1997 Edition (in Japanese).

58 U.S. Department of Commerce. ibid. p.48

59 U.S. Department of Commerce. ibid. Appendix 4, p.37

60 U.S. Department of Commerce. ibid. Appendix 4, p.37

61 Here it is necessary to distinguish between development and growth. Kindleberger offers the following: "Economic growth means more output, and economic development implies both more output and changes in the technical and institutional arrangements by which it is produced" (Kindleberger, C.P. Economic Development, 2nd Edition, McGraw-Hill, 1965. p. 3).

62 Nishikawa, J. *Keizai hatten no riron* (Theory of Economic Development), Nihon Hyoron-sha, 1978 (in Japanese). p.22.

63 *China Daily*. "China: Boosting IT Industry to Meet World Trend," April 6, 1998.

64 Murota, Y. and Itoh, K. "Global Warming and Developing Countries," *Energy Policy*, vol. 24, no. 12, 1996.

65 U.S. Department of Commerce. ibid. p.49

66 Nishikawa, J. ibid. p.233

67 For example, C. Whitmer, who was in charge of NT graphics at Microsoft, is described as follows: "To Whitmer, freedom meant more than refusing to file status reports, attend meetings or answer the phone...Freedom meant something more profoundly liberating to Whitmer, who promptly declared himself "dead" and disappeared. His team disappeared with him...unfettered by the ordinary rhythms of bureaucracy...called his team the Undead." (G. P. Zachary. Showstopper!: the breakneck race to create Windows NT and the next generation at Microsoft. The Free Press. 1994.)

68 The background of deregulation is explained as follows (Ide, H. "*Chokusetsu kisei seisaku*" (Direct Regulation Policy), in Sinjo K., ed., *Sangyo soshiki-ron* (Theory of Industrial Organization), Yuhikaku Books, 1995.):

1) The need for regulations

- In areas with natural monopolies and uneven distribution of information, in order to prevent inefficient resource allocation and to protect consumers, governments protect consumers by regulating entry and withdrawal of companies, price, service quality and quantity, etc.
- Examples of natural monopolies: Electric power, communications. Examples of unequal information distribution: Finance, etc.
- That is to say, governments intervene in cases of market failure. With natural monopolies it is decreasing cost industries, and with finance it is incompleteness of information, etc.

2) Background of deregulation (as above)

- There are "failures of government."

a. Regulating authorities have difficulty obtaining the information needed to control regulated companies.

b. Regulation is influenced by corporate interests and politics, and therefore tends to reflect the preferences of the regulated companies (capture theory).
- The existence of regulation brings about rent in existing industries and companies.

69 U.S. Department of Commerce. ibid. Appendix 4, p.47

70 U.S. Department of Commerce. ibid. Appendix 5, p.11

Chapter 5

THE ROLE OF INFORMATION DISCLOSURE IN CORPORATE GOVERNANCE: THE CASE OF THE CHEMICAL INDUSTRY

CINDY TERMORSHUIZEN
IGES (Institute for Global Environmental Strategies)

1. INTRODUCTION

The corporate world is in an extraordinary state of flux. A combination of rapid technological change and economic liberalization have revolutionized 'old' sectors like finance and telecommunications, kept up the pace of change in the always dynamic computer industry and given birth to whole new areas of activity like biotechnology. These changes have led to the proliferation of new companies and the rapid growth of many old ones. This increasingly complex and powerful corporate world finances most of the world's economic activities, creates the vast majority of its pollution and trains and employs a large share of its workforce.

With the growth of corporations, both in size and numbers, has come an increasing need to govern them. At the same time, however, the sheer scale and complexity of the corporate world often renders traditional governance mechanisms inadequate. Technology-specific regulations, for instance, tend no longer to make sense in a world of rapid technological change.

The governance challenges posed by the corporate world are perhaps most pronounced in the chemical sector. The world's chemical companies contribute U.S.$1.5 trillion to the global economy and hundreds of thousands of jobs. They also add 1000 new chemicals annually to the now over 85,000 chemicals already in production. For many of these chemicals, only time will tell if they are environmentally benign or not. However, the devastating environmental and human

K. Matsushita (ed.), Environment in the 21st Century and New Development Patterns, 115–145.

health effects of chemicals such as DDT, PCBs, and dioxins have drilled home the need for constant vigilance.

The value of the chemical industry to the global economy, on the one hand, and the potential risks of its products, on the other, require a balancing act on the part of corporations and their stakeholders: government, the financial sector, labour and consumers. Stakeholders either benefit from corporate financial success or suffer the adverse effects of inefficiency and mismanagement; either way, they need to monitor companies' achievement of performance goals and compliance with regulations. This balancing act is what corporate governance[1] is all about, and the information used for monitoring corporate activities is its most potent tool.

In order to satisfy the expectation that chemical corporations be both financially successful and environmentally responsible, corporate governance must center on the promotion of innovation, competitiveness, and good environmental performance. The most powerful tool to achieve these goals is well-designed information disclosure rules. While information disclosure is important in all sectors of the economy, it is absolutely vital in sectors -- like chemicals -- that are characterized by rapid product development and high levels of complexity. Straightforward command-and-control regulation does not function well in such environments.

This paper takes a closer look at the role of information disclosure in the governance of chemical corporations. After a brief discussion of the goals of corporate governance in general, the paper goes into an analysis of the role of information in corporate governance. It argues that, while information disclosure mechanisms are increasingly widespread, a failure to develop and implement them systemically renders them only partially effective. To illustrate this point, the paper goes on to look at the mechanisms of corporate governance prominent in the chemical industry, the information disclosure components they include, and the extent to which these promote innovation, competition and better environmental performance. The paper closes with a set of recommendations for strengthening information disclosure in corporate governance mechanisms.

2. THE GOALS OF CORPORATE GOVERNANCE

While the overriding goal of corporate governance is a balancing of the corporate need to be financially successful and the public's and environment's need to be protected from harm in that process, the underlying goals essential to achieving this are: the promotion of innovation, competitiveness and superior environmental performance on the part of corporations. Each of these is examined in more detail in the following paragraphs.

The first-two goals of corporate governance, innovation and competitiveness, are closely linked. Simply stated, the most innovative sectors of modern economies are the most competitive, because innovation means new processes, new products, new markets and -- as a result -- higher corporate revenues (Day 1998).[2] Given the rapid pace of technological change in key areas of modern economies, such as finance, telecommunications, and biotechnology, innovation is central to the survival of firms.

Innovation and competitiveness largely satisfy the market side of the corporate governance equation, but environmental considerations also need to be taken into account. Most broadly, the environment can be taken to mean both the social and natural environment. For the purposes of this paper, however, the focus is on the latter, particularly the area of resource efficiency and waste management. Of course, poor corporate environmental performance often has dire social implications. Poor natural resource management, for instance in the area of fisheries and forestry, may endanger the livelihood of local people. Poor waste management on the part of firms endangers worker and community health and safety.

Because poor natural resource and waste management also translate into lower material efficiency, higher clean-up costs and lost sales and investment due to consumers' and investors' negative responses, environmental considerations are in fact an integral part of the innovative and competitive part of the governance equation. Thus, energetically tackling environmental performance issues is central to corporate governance.

3. THE ROLE OF INFORMATION IN CORPORATE GOVERNANCE

While command-and-control style regulation, defined by its heavy emphasis on emissions standards and licenses, has been and continues to be the major tool used by governments to govern corporate behaviour, particularly in the area of pollution control, different types of regulation as well as a whole set of voluntary measures are becoming increasingly important.[3] At the heart of many of these measures is an information disclosure component. Though this paper focuses on how it is used in regulatory and voluntary mechanisms for efficiency improvement and waste reduction in the chemical industry, information disclosure is being used in all areas of corporate governance.[4]

Disclosure requirements aimed at corporations can serve a number of purposes. At their best, they are a tool to help companies understand their own resource profiles better, leading to innovation, improved resource efficiency, decreased waste and the enhancement of competitiveness. While it is often assumed that corporate managers already have a complete understanding of their resource consumption patterns and opportunities for innovation, case study results indicate that this is not so. In reality, managers "often have highly incomplete information and limited time and attention" (Porter and van der Linde 1995:127). Regulatory and voluntary mechanisms sometimes include reporting requirements to address these problems.[5]

A second possible purpose of disclosure requirements is to aid governments in both monitoring the effectiveness of any pollution reduction programs that they might have in place and in setting priorities for future action. Disclosure requirements with this purpose also have direct implications for corporate governance, as they reveal whether or not existing programs are shifting corporate attention toward innovative solutions to waste problems.

Often the purpose of disclosure requirements is more general. The premise upon which many are based is simply that information disclosure is necessary because public, government and investor pressure are the key impetus behind improvements in corporate environmental performance. This pressure cannot be effective, the argument goes, if it is not based on reliable, detailed information. The flaw in many disclosure requirements based on this premise is a failure to take into consideration the need both to provide access to disclosed information and to train people to understand and use it. As will be discussed in subsequent sections of this paper, this flaw has particularly serious consequences in those developing countries where access to information (due to infrastructural problems) is limited and literacy rates low.

4. MECHANISMS OF CORPORATE GOVERNANCE

Given the growing importance of information disclosure requirements in corporate governance mechanisms, the question arises as to how effective these requirements are in promoting innovation, competitiveness and superior environmental performance. This section briefly outlines what the major corporate governance mechanisms are and then goes into a deeper discussion of the ways in which they promote information disclosure and the extent to which that information is or can be used to enhance innovation, competitiveness and environmental performance. Throughout, examples from the chemical industry are used. The findings of this section suggest that a combination of poor design and systemic impediments can keep disclosed information from contributing in a substantial way to improving innovation, competitiveness and environmental performance in the corporate world.

OVERVIEW OF GOVERNANCE MECHANISMS

The key mechanisms of corporate governance relevant to resource efficiency and waste management can be divided into regulatory, voluntary and market mechanisms. The three overlap considerably in practice, particularly in some countries.[6] However, they will be addressed separately for the purposes of analysis.

Regulatory mechanisms refer to those sets of rules enshrined in law, administered by governments, and backed up by the threat of legal sanction. They vary widely between countries and over time. For instance, in the process of developing rules, governments may or may not consult the relevant industries for input. As for the actual shape of the rules, they can be technology specific or results oriented and may or may not rely on market incentives, such as pollution charges and tradable permits. Regulatory mechanisms at the international level share many of the traits of national level regulation. However, due to systemic constraints, international law tends to be much weaker than national law.

Voluntary mechanisms are non-regulatory.[7] Sometimes they are unilateral commitments made by a particular firm. Usually they take the shape of voluntary agreements among groups of corporations, or between corporations and government and/or non-governmental organizations (NGOs). Increasingly, voluntary agreements are being struck between corporations and NGOs without any substantive government involvement. All of these mechanisms can and do operate at both the national and international level.

Finally, there are market-related mechanisms. For the purposes of this paper, these refer to non-regulatory initiatives taken by banks, insurance companies and investors to monitor and govern corporate behaviour in the area of resource efficiency and waste management. Because these initiatives are backed up by market incentives, they can be extremely effective.

4.1 Regulatory Mechanisms at the Government Level

While the focus of this paper is on information disclosure in the context of waste management and resource efficiency, no discussion of regulatory mechanisms in this area would be complete without mention of basic access to information legislation. This type of legislation exists in some of the advanced industrialized countries, and campaigns to introduce it are being waged in a number of industrialized and industrializing countries that as yet have no such legislation.[8] Laws enshrining citizens' right to know can provide a context in which information disclosure is encouraged. In practice, however, provisions to protect trade secrets can act as an impediment to the disclosure of corporate information.[9] That said, evidence from the United States suggests that chemical companies rarely attempt to classify chemical use data as trade secrets (Ranganathan 1998:4).

Access to information laws have been of critical importance for several decades in informing citizens and investors about corporate environmental performance in some countries. However, the specific inclusion of extensive information disclosure stipulations in corporate resource efficiency and waste management related regulations is a much newer phenomenon. Some of the most comprehensive legislation in this category is the so-called Pollutant Release and Transfer Register (PRTR).[10] A Pollutant Release and Transfer Register is a "catalogue or register of potentially harmful pollutant releases or transfers to the environment from a variety of sources" (OECD(a):15). Individual governments establish which pollutants to classify as "potentially harmful" and which sources (mainly industrial sites) will be required to report on them. PRTRs are generally made available to the public, excluding only that information which has been established as trade secrets.

The OECD argues that the information disclosed in PRTRs promotes pollution prevention for three reasons. First, it assists government authorities in setting priorities for reducing or eliminating the most harmful pollutants. Second, it prompts generators of pollution to re-evaluate their production processes by showing the extent of resources lost as waste. Finally, it instills competition among pollution generators based on the fear of being labelled as environmentally irresponsible by the general public (OECD(a):15-16).

A convincing amount of evidence exists to support this claim. In both the U.S. and Canada, which have had PRTRs in place since 1986 and 1993 respectively, significant reductions in pollutant releases to the environment have been documented. According to the United States Environmental Protection Agency, between 1988 and 1995 releases of designated pollutants to the air decreased 46.1 percent, to surface water 78.2 percent, to land 42.2 percent and to areas underground (by underground injection) 15.6 percent (EPA(a)). In Canada, between 1995 and 1996 alone, on site releases declined 14.9 percent (Environment Canada 1996).

Industry spokespeople support the idea that these reductions are related to the establishment of PRTRs. For instance, the chief executive officer of Du Pont Corporation, the world's largest chemical company, claims that Du Pont's position as number one polluter on the 1988 Toxic Release Inventory (the U.S. PRTR) led the company to take a more proactive position on emissions reduction and pollution prevention, even though all of its emissions at the time were legal. The rationale behind this move was simple; the CEO noted that "given the tenor of the times and the mood of the public, these emissions would not be tolerated indefinitely" (Smart 1992:187). While perhaps not entirely attributable to the Toxic Release Inventory, Du Pont reduced its total waste generated by an astonishing 46 percent between 1991 to 1997 (Du Pont 1997).

Though significant decreases in pollutant releases to the environment have been documented following the introduction of PRTRs, they have often not been accompanied by resource efficiency innovation and the higher levels of competitiveness that accompany such innovations. For example, U.S. statistics for 1995 to 1996 show that while pollutant releases dropped by 4 percent, the amount of production-related waste managed by industry rose by 1.9 percent (PIRG(c) 1998). During the same period in Canada, off-site transfers of waste rose 24 percent (Environment Canada 1996). This is because many companies have relied on end-of-pipe controls like recycling rather than process innovation to reduce pollutant releases (EDF(a) 1998). Shifts in the kind of information reported, the shape of the legislative and bureaucratic systems in which pollution prevention activities take place, and the attention paid to citizens in the process could greatly strengthen PRTR laws and lead to higher levels of innovation, competition and environmental performance.

First, a shift from reporting transfers and releases only to reporting how chemicals are actually used in the production process and in products themselves would shift manufacturers' focus from waste management to process innovation. As Kevin Mills, director of the Environmental Defense Fund's Pollution Prevention Alliance notes, "[t]o put pollution prevention first, right-to-know programs must be expanded to track toxins in manufactured products, within manufacturing processes, and in upstream activities such as material processing and transportation" (EDF(a) 1998). The American states of New Jersey and Massachusetts have legislation in place which has begun to expand information disclosure programs in this way and requires corporations to report on toxic chemical use. The results indicate significant cost savings and emission reductions. For instance, a Massachusetts Department of Environmental Protection study revealed that between 1990 and 1997, Massachusetts industries saved $14 million while reducing toxic chemical use by 20 percent and toxic waste generation by 30 percent (PIRG(b) 1998).

For such expanded information disclosure laws and programs to result in increased innovation and competitiveness and better environmental performance, legal and bureaucratic systems will need to be reoriented to these goals as well. Present systems often stifle the innovation that process-oriented information disclosure could spur. The example of India is instructive. While industries in that country are required to submit detailed reports identifying both the sources of pollution and any control measures in place, tax and subsidy structures fail to provide industry with incentives to choose least cost pollution abatement technologies (Murty 1995:331-333), to say nothing of choosing process technology that would reduce pollution at the source. Where taxes and subsidies are not a problem, technology specific legislation often is. Laws requiring the use of so-called 'best available technology' imply that one technology is 'best' and discourage the search for innovative, and perhaps even better, alternatives (Porter and van der Linde 1995:124). Legal problems are compounded, and often created, by bureaucracies that are structured with a focus on fragmented, end-of-pipe solutions to waste. Attempts to re-orient them toward holistic, process-based approaches to waste management face much resistance from within affected agencies (Burnett 1998).

Finally, if PRTRs are to live up to their potential, it is vital that citizens be capable of accessing and using them. Public opinion can be a powerful incentive for firms to innovate and improve their environmental and safety records, but local citizens are unlikely to effectively pressure firms in their neighborhoods if they cannot access or do not understand the data which the law makes available to them. Countries like the United States, with strong environmental groups[11] and Internet access at many public schools and libraries, have a head start in this area. Other countries that have weaker environmental groups and less access to information must recognize that PRTRs lose a great deal of their effectiveness under these conditions. The placement of documents in libraries or on the Internet and the provision of training for citizens' groups could go far in remedying these problems.

In short, information disclosure requirements embedded in law can be effective tools for the promotion of innovation, competitiveness and improved environmental performance in the chemical and other industries. Regarding the chemical industry, particularly in cases where both general right to know laws and PRTRs are in place, declines in chemical releases have been dramatic in recent years. That said, the failure of most PRTRs to focus on chemical use (as opposed to release) has led to a continued focus on end-of-pipe solutions to pollution problems. The release of use information in easily understood formats is a more effective means of promoting both innovation and improved environmental performance in the chemical industry.

4.2 Regulatory Mechanisms at the International Level

Information disclosure is also becoming an increasingly important aspect of international law. Most international law relating to the environment, and particularly waste management, attaches importance to "procedural obligations of ...monitoring, notification, and consultation with other states and international institutions"(Birnie and Boyle 1992:303-304). However, even in highly regarded agreements such as the Basel Convention, disclosure rules are rendered less effective due to low levels of "transparency and public access to decision making... (and) the absence of any provision for independent inspection" (Birnie and Boyle 1992:342). Furthermore, the usefulness of international agreements as tools for corporate governance and the promotion of innovation, competitiveness and superior environmental performance is limited by the highly state-centric nature of rule-making at the international level.

A good example of both the promise and problems of information disclosure at the international level is the Rotterdam Convention on the Prior Informed Consent Procedure for Certain Hazardous Chemicals and Pesticides in International Trade, usually referred to simply as the PIC Convention.[12] The purpose of this legally binding Convention is to provide information to governments on a set of specified chemicals (presently 22 agricultural chemicals and 5 industrial chemicals) and allow them to decide whether or not to ban the import of these substances. It also leaves compliance measures up to national governments.

Because the PIC Convention is not a recommendation to ban or restrict the production or use of hazardous chemicals per se, its implications for corporate governance in the area of environmental performance are not clear. Nonetheless, the complexity of the procedure to add new chemicals to the Convention,[13] the difficulties faced by developing countries in effectively using information provided by the Convention, and the failure of the Convention to include provisions for information disclosure by the agrochemical industry suggest that it will fail to encourage chemical companies to innovate and produce less environmentally harmful products.

For the time being, the filling of the gaps in the Convention may largely be up to non-governmental organizations (NGOs) and universities, both in industrialized and developing countries. For one, these groups could provide training for developing country customs officials, agricultural extension and health workers in accessing and using PIC data and recognizing the health effects of hazardous chemicals.[14] They could also lobby agrochemical corporations to release data on potentially hazardous chemicals. The latter would have to be done largely by NGOs and academics in the home countries of chemical producers, as disclosure laws are weak or non-existent in many of the developing countries where production takes place. However, it is important for developing country NGOs and universities to pressure chemical corporations in their countries to disclose information in a systematic fashion. This is necessary because, increasingly, production is taking place in developing countries. As many chemical corporations in developing countries are branch plants of multinational corporations based in advanced industrialized countries, these efforts could be made easier were links established between developing and industrialized country NGOs and academics in this issue area. These efforts could greatly enhance the effectiveness of the PIC Convention, particularly if they receive solid media coverage.

While NGOs and universities can do a great deal to press for innovation and superior environmental performance on the part of chemical corporations, the question arises as to why this should be necessary at all. Why is the Convention (and many other pieces of international law similar to it in terms of information disclosure) not sufficient? Perhaps the failure of the Convention to act as a force for innovation and enhanced environmental performance on the part of chemical corporations can be attributed to its failure to approach the issue of hazardous chemicals from an efficiency perspective. In other words, the framers of the Convention did not see the negative environmental performance of hazardous chemicals as evidence of inefficient resource use and thus did not write the Convention in such a way as to promote innovation in this area.

An example from the PIC Convention illustrates this point. One category of chemicals that is eligible to fall under the Convention is the so-called "hazardous pesticide formulations". To such chemicals, strict application restrictions are applied that include "closed tractor cabs or full body respirators" (USIA 1998). The high levels of toxicity suggested by these measures mean that large amounts of resources (human and otherwise) must go into just transporting and using these chemicals. In most cases, disposal of any unused chemicals in this category would be resource intensive, as would the clean-up of accidental spills. If hazardous chemicals are viewed in this light, they are highly inefficient ways of getting rid of agricultural pests and their production and use should be discouraged. Were an efficiency perspective of this sort to underlay the PIC Convention, procedures for adding chemicals to the Convention would be greatly simplified and resources would be allocated to ensure access to and effective use of the information flow it unleashes.

In conclusion, while international legal mechanisms like the PIC Convention are overly state-centric and end-of-pipe in approach, thus failing to encourage efficient resource use by corporations, they have the potential to unleash a great deal of useful information. Through the efforts of NGOs and academics in particular, this information can be used to promote innovations in resource efficiency and waste management, thus improving corporate competitiveness. In the case of the PIC Convention, for instance, use of released information by NGOs and academics may speed inclusion in the Convention of additional hazardous chemicals and force compliance by national governments. These activities, in turn, could induce agrochemical companies to focus their research on developing less hazardous chemicals.

4.3 Voluntary Mechanisms

Along with the slow shift away from command and control regulation at both the national and international levels has come an increasing emphasis on voluntary mechanisms for reducing industrial emissions. There are essentially three types of voluntary mechanisms: unilateral commitments, public voluntary schemes and negotiated agreements.[15] Unilateral agreements are undertaken by corporations, either alone or as a group. Public voluntary schemes are designed by governments and voluntarily joined by corporations. Finally, negotiated agreements are drawn up by corporations in cooperation with governments and/or NGOs. Many of these voluntary mechanisms have significant information disclosure components.

4.3.1 Unilateral Commitments

Unilateral Commitments by Individual Firms

The most important voluntary information disclosure activity by individual firms is the release of environmental health and safety (EHS) reports. The number of corporations preparing such reports has increased markedly in the past five years. EHS reports have a variety of aims which may include assuring host communities that plants located in their areas are safe, convincing environmental groups and employees that they are making strides to improve their environmental performance, and showing investors that environmental liabilities do not exceed corporate abilities to handle them (Davis-Walling and Batterman 1997:865). The aim of the report significantly affects the nature of its contents.

A range of opinion exists on the value of the information in these reports, particularly to investors. Crossley and Points of the World Wide Fund for Nature (WWF) argue that the bulk of investors see environmental health and safety reports as largely qualitative documents that fail to make a clear link between a company's environmental spending and its profitability (p.8). Alison Cassidy of the Investor Responsibility Research Center does not refute this argument. However, she notes that the EHS reports of major American multinationals can be a rich source of data on corporate operations, particularly those operations that occur outside of the United States and are not covered by other information disclosure mechanisms.

While these reports may be useful to investors and local communities, a number of weaknesses in the reporting process tend to hinder their use as truly effective tools for corporate governance. The first problem is selective reporting. Because no quality-control system exists to ensure that the information contained in EHS reports is comprehensive, many companies choose to report only that information which enhances their public image (Davis-Walling and Batterman:875). In addition, most firms fail to report on environmental aspects of their overseas operations, an important omission given that many of the corporations in high impact industries, like fossil fuels and chemicals, do a significant part of their productive activities abroad.[16] This does not automatically mean that a corporation is environmentally irresponsible, but it does make it difficult for the corporation to use the report as a springboard for innovation and for investors and NGOs to use it to pressure the corporation, either financially or morally, to innovate.

A second, and related, problem is that information in environmental reports is difficult to compare (Crossley and Points:8). Though there are a number of guidelines that help corporations draw up reports,[17] none of these contains measures which formally compare the environmental spending or liabilities of a particular firm to those of other firms or the industry average (Davis-Walling and Batterman 1998:870). This weakness is of particular concern to investors.

A final problem is that smaller corporations tend not to put out environmental reports (Davis-Walling and Batterman:873). While this is understandable given the expense involved in compiling an environmental report, it is worrisome, particularly in some industries where the contribution of small firms to overall pollution is significant. For instance, the Environment Agency of Japan reports that small plating shops in that country are major emitters of toxic waste, as are dry cleaners and hair dressers (The Economist(a) 1998).

In short, mechanisms are needed to ensure that voluntary reporting by individual firms is comprehensive, comparable and accurate. In industries where small firms are significant producers of toxic emissions, some financial and technical support will probably be necessary if these small firms are to be induced to report voluntarily on their emissions. Incentives, such as positive publicity for reporters, may also induce firms to voluntarily disclose information.

Unilateral Commitments by Groups of Firms

In addition to voluntary activities undertaken by individual firms, many cooperative arrangements exist among firms within particular industries. Often these are coordinated by trade associations. One of the most important of these, both in terms of its scope and the demonstration effect it has had, is the chemical industry's Responsible Care program.[18] It was launched by the Canadian Chemical Producers Association (CCPA) in 1985 and has spread to over 40 countries, accounting for 86 percent of world chemical production (by volume) (Mazurek1998:5). Under Responsible Care, geared more toward changing management practices than developing environmental performance standards, member companies agree to continuously improve their environmental, health and safety performance and strengthen their public outreach (Mazurek 1998:6). The program's underlying rationale is that the improvement of the chemical industry's public image, battered by numerous accidents, is critical to its survival (Mazurek1998; Borkey and Leveque 1998). This is so in part because permits to expand operations are difficult to obtain if local residents and environmental NGOs vigorously oppose them. Tougher national level legislation is also more likely if public opinion is strongly negative.

The information disclosure mechanisms incorporated into Responsible Care vary depending on the producers association through which the program is implemented and have important implications for both waste reduction achievements and public support of the industry. The most important differences between programs concern verification mechanisms and target setting. A comparison of Canada and the United States in these two areas is instructive. While the Canadian Chemical Producers Association (CCPA) has in place a reporting verification process which all members of the Responsible Care program have been required to use since 1993, the verification system of the U.S. Chemical Manufacturers Association (CMA) is optional. At present, only 10 percent of the CMA's Responsible Care members use it, despite the fact that participants are not required to disclose their reports to the general public (Mazurek 1998:9). The CCPA also differs from the CMA in terms of target setting. While the former has set an emission reduction target for the Association as a whole,[19] the CMA has set no targets.

Data on the emission reductions achieved and public support enjoyed by the Canadian and U.S. chemical associations suggests that weak information disclosure systems, such as that of the CMA, lead to lower waste reduction and lower public support than that observed where strong, results-oriented information disclosure systems are in place. The CMA's failure to put in place a strong verification system and clear reduction targets means that only anecdotal and case-specific data exists on the effectiveness of the program.[20] This data is mostly positive- pointing to "lower costs from treatment of injuries, reduced losses of materials in spills, and improvements in regulatory compliance" (Mazurek 1998:17)- but is widely mistrusted, particularly by environmental NGOs.[21]

If enhanced public opinion is crucial to the chemical industry's efficient operation and very survival, stronger disclosure mechanisms- jointly devised by industry and concerned environmental, community and labour groups- appear to be in order. There is an important precedent within Responsible Care for such a cooperative approach. Perhaps the most successful part of Responsible Care in the United States has been its community advisory panels (CAPs); these appear to have built trust in local communities through meetings between residents and plant managers of local plants regarding plant expansion and environmental improvement plans (Mazurek 1998:14). In surveys, people living in the vicinity of plants and local politicians who are familiar with the Responsible Care program have more positive opinions of industry than those individuals who are unaware of the program (Mazurek 1998:18). Extending the participatory, information-sharing nature of the CAPs to the Responsible Care program as a whole could greatly improve public trust in the program. This, in turn, could make chemical companies more competitive, as higher consumer confidence would reduce the threat of consumer boycotts, increase sales, and increase banks' willingness to extend loans.

While strengthening the information disclosure and participatory elements of Responsible Care programs makes sense from an efficiency perspective, legal obstacles to such actions may exist. The most important of these is the persistence of a patchwork of media-specific (air, water, land) laws in many countries which overlap with the more integrated, prevention-oriented focus of voluntary programs like Responsible Care (Mazurek 1998:5). Fulfilling the reporting requirements of both is irksome to industry and may not be all that useful to regulators, investors and the public either. The piecemeal data that results from these multiple reporting requirements makes it difficult for stakeholders and corporations themselves to form a clear picture of their performance in relation to other companies. A rethinking of reporting requirements for industry appears to be in order.

Anti-trust legislation, a factor in approximately 80 countries world-wide, may compound these problems. Such legislation, which is aimed at preventing corporations from gaining monopoly power in an industry limits the extent to which trade associations may employ sanctions to force compliance with reporting and other measures. The existence of anti-trust legislation in the United States has arguably hindered the ability of the Chemical Manufacturers Association to enforce reporting requirements (Mazurek 1998:5). Given the importance of both anti-trust legislation[22] and information disclosure in promoting competitiveness, the case for a more comprehensive regulatory structure for information disclosure and third party verification becomes considerably stronger.

4.3.2 Public Voluntary Schemes

One possible way to begin to overcome the reporting problems associated with unilateral voluntary programs like Responsible Care is through public voluntary schemes. As explained above, public voluntary schemes are designed by government and joined voluntarily by industry, though often industry participation is 'forced' by the threat of regulation or by public and peer pressure. The most well known and extensive schemes in this category are the global ISO 14001 and the European Eco-Management Auditing Scheme (EMAS). Both of these are environmental management systems designed to provide participating companies with a framework for improving their environmental performance.

There is both a direct and an indirect link between these programs and the enhancement of competitiveness. Most directly, many corporations and government bodies are already, or are now considering, requiring their suppliers to have ISO

14001 or EMAS certification. It appears that having certification in one or both of these programs is becoming a prerequisite of doing business in many parts of the world.[23] More indirectly, the auditing process built into both of these programs provides corporations with information on how to improve their environmental performance. By systematically checking that their operations comply with all regulations, companies can reduce the risk of heavy fines and costly legal battles as well as reduce their insurance rates (Bruner and Burns 1998:18). The publication of information related to these audits can also assist governments, investors and the public in identifying which companies have good environmental records and which have considerable room for improvement.

The approaches of ISO 14001 and EMAS to information disclosure have significant differences, with the latter placing much greater emphasis on making information available to the public. To be accredited under ISO 14001, companies must show that their in-house environmental management system (EMS) measures up to that of ISO 14001. Once revised, the EMS must be implemented and continuously monitored, and corporate performance must be audited (Bruner and Burns 1998:18). However, only the EMS itself need be made public. Under the Eco-Management Audit Scheme (EMAS), a great deal more information is disclosed. An environmental statement, validated by accredited environmental verifiers, is publicized which includes "a description of the site's activities; an assessment of all the significant environmental issues; a summary of figures on pollution emissions, waste production, consumption of raw material, energy and water, and noise; a presentation of the company's environmental policy and site's programme and management system; the deadline for the next statement; and, the name of the accredited environmental verifier"(EMAS 1998).

It could be argued that the ISO 14001 standard would benefit from adopting the EMAS environmental statement. Of course, given the relative newness of both ISO 14001 and EMAS, there is no empirical evidence to suggest that the fuller disclosure provided by EMAS leads to superior environmental performance than that which would occur under the ISO system. Nonetheless, information disclosure evidence from regulatory and other voluntary programs shows that the more extensive disclosure is in a program, the more trust it gains from investors and the public and the more incentive there is for corporations to innovate and reduce waste.

Though there is room for improvement in the ISO 14001 environmental management system, its positive contribution to environmental information disclosure on a global scale is undeniable. ISO 14001 has been particularly useful to many developing countries in providing a blueprint for national corporate disclosure rules. The Mexican Environment Ministry has added a voluntary EMS to its regulatory framework, and both the Philippines and Indonesia plan to incorporate elements of ISO 14001 into new programs for corporate environmental information disclosure (World Bank 1998). However, firms in many other developing countries do not have the capacity to apply for ISO certification. In order for corporate information disclosure not to become a further handicap to developing country industries, capacity building efforts in this area will need a great deal of attention.[24]

4.3.3 Negotiated Schemes

The final type of voluntary scheme discussed in this paper is the negotiated scheme. Negotiated schemes, as the name implies, are the result of negotiation between industry and government and/or non-governmental organizations. Negotiated schemes between industry and government are by far the most common, but NGO involvement is becoming an increasingly important factor.[25] The various types of negotiated schemes and their information disclosure components are discussed below.

Negotiated schemes between industry and government are by no means new. In Japan, for instance, they have existed since the 1960s. Presently there are more than 30,000 of them (Imura 1998:2).[26] In the European Union they are smaller in number but growing. A 1996 survey revealed that over 300 negotiated agreements are recognized by EU country governments (European Environment Agency 1997:3)

While information disclosure mechanisms vary significantly across these schemes, a number of common concerns exist. The most important of these are the transparency of the negotiation process and the status of the information released as a result of the scheme. The latter is of particular concern when a semi-governmental body is set up by government in order to conduct negotiations with industry and perform certain policy functions. In this case, conflict can arise over the status of information generated by this body and whether it is accessible to the public and other parties not involved in the negotiations (European Environment Agency 1997). Situations like this can seriously undermine the credibility of negotiated agreements and need to be avoided at all costs.

One way of both increasing the transparency of negotiated agreements and ensuring that the information generated remains in the public domain is to involve NGOs in the negotiation process. Canada's Accelerated Reduction/Elimination of Toxics (ARET) Program is one scheme where this occurred. Government, professional, health, environmental, labour and industry representatives formed the committee that decided which chemicals (117 in total) to target for reduction or elimination (Riordan:46). Though environmental and labour groups eventually withdrew from the process due to disagreements with the rest of the committee about which substances to eliminate (as opposed to reduce), their involvement set a precedent for openness, both at the negotiation and later stages. ARET has a number of weaknesses, such as a lack of third party verification for reporting, but its involvement of all stakeholder groups and the great interest from industry[27] which it drew despite (or because of) this is indicative of the potential of programs involving all stakeholders.

The question of whether or not negotiated schemes promote innovation and thus competitiveness and improved environmental performance has not yet been answered. Opinions on this matter remain mixed. Some commentators argue that in the negotiation process, collective learning takes place and innovation is spurred. Others argue that, among other things, the unambitious targets of negotiated agreements lead to lower innovation that would be seen were taxes or other market-based approaches taken (Borkey and Leveque 1998:26). While both of these verdicts are probably true depending on the voluntary scheme in question, one of the most important factors determining the level of innovation resulting from these agreements is information. If source-oriented, specific and comparable information is released, there will be greater incentive for the setting of ambitious targets and compliance with these through process innovation.

4.4 Market-based Mechanisms

Regulatory and voluntary mechanisms play a crucial role in governing corporate behaviour. However, investors, banks and insurance companies play an equally important role. Without their confidence and the money which flows into firms as a result, modern firms simply cannot exist. Publicly quoted corporations are particularly vulnerable, as their performance can be judged daily on the stock market. This section briefly looks at the rising interest on the part of banks, insurance companies and investors in obtaining corporate environmental information, the special role of third party rating agencies and other financial research institutions in providing the desired information, and the implications of these actors' activities for corporate innovation, competitiveness and environmental performance.

In both banks and insurance companies, there is a growing awareness that poor corporate environmental performance has important links to management and financial performance. For bankers, any environmental cleanup obligations may reduce firms' ability to pay off loans. As well, property used as collateral by a borrower may decline in value if contaminated (Labatt and Maclaren 1998:201). Such firm-specific issues matter to insurance companies too, though increasingly global environmental problems are cause for concern. In 1995, insurance companies paid out over $110 billion for damage from natural causes, three times the amount paid out in the previous year (BMU). The blame for much of the damage was attributed to climate change, due largely to human activities like the combustion of fossil fuels.

As a result, banks and insurance companies worldwide are demanding sound environmental performance from their clients. The level of commitment to this is indicated by the great interest being shown by the banking and insurance sectors in the UNEP Statement by Financial Institutions on the Environment and Sustainable Development and the Statement of Environmental Commitment by the Insurance Industry. More than 100 financial institutions have signed the former, and more than 70 the latter. These statements commit signatories to not only review their own in-house environmental performance but also to pay greater attention to the environmental performance of their clients when considering loans and drawing up policies. Signatories to the Statement by Financial Institutions, for instance, pledge their support for "the precautionary approach to environmental management" and note that they "regard compliance with applicable environmental regulations and the use of sound environmental practices as important factors in demonstrating effective corporate management"(UNEP(a) 1997).

Individual and institutional investors, too, are increasingly concerned about the environmental performance of the firms in their portfolios. One indicator of this growing concern is the amount of assets now being managed in socially and environmentally responsible (green) funds. The Social Investment Forum reports that, in the U.S. alone, over $1 trillion in assets are being managed in such funds. That is almost double the amount in 1995 (IRRC). However, even outside of specifically green funds, environmental performance matters. In both industrialized and developing countries, investors are quick to withdraw their funds from companies revealed to have poorer than expected environmental performance, thus reducing share values (Konar and Cohen 1997;Dasgupta, Laplante and Mamingi 1997). The opposite is also true; research suggests that firms with better environmental performance outperform their competitors on the stock market (Konar and Cohen 1997).

The combined interest of banks, insurers and investors in environmental performance is one of the driving forces behind the trend toward greater corporate information disclosure. Experts in the field feel that the growth of environmentally responsible funds, in particular, will increasingly lead investors to insist on compliance with disclosure laws (Campanale 1994:43). Their insistence will be backed up by the efforts of researchers at green funds and at environmental rating agencies.

Environmental rating agencies are of special value in the quest for better corporate information disclosure. These agencies, usually funded and utilized by investors, assess corporate environmental risks and returns. Dirk Reinhard of O kom, a German rating agency, argues that companies are essentially compelled to have a rating because investors demand it (Wong 1998). These efforts are supplemented by researchers at individual green funds. They generally must rely on existing information supplied by corporations, but do frequently contact these corporations to request fuller disclosure on relevant issues (Campanale 1994:44). Though the focus of much of the rating and research activity tends to be on environmental liabilities rather than assets, some funds undertake in-depth process and product analysis to get a picture of both the positive and negative aspects of a firm's environmental performance (Campanale 1994:45).

While little data exists to determine the impact of increased emphasis by the financial community on corporate environmental performance and information disclosure, openness about environmental performance appears to play a part in spurring innovation and improving environmental performance in the long run.

Dupont, for instance, releases detailed environmental financial reports and notes that its environmental impacts declined 50 percent while shareholder value rose 200 percent between 1989 to 1997 (Wong 1998).

Numerous studies of changes in stock prices after media reports and PRTR filings show that firms that lose the most money subsequently innovate the most to improve their performance. In short, information disclosure and reactions to that information by the financial community appear to be a key factor in promoting waste reduction and innovation.

5. THE ROLE OF THE MEDIA

A discussion of the various types of disclosure mechanisms relevant to corporate environmental performance is not complete without mention of the media. As alluded to in the above pages, media reports both complement other reporting mechanisms and compensate for them where such mechanisms do not exist. In the former case, the media may report on data that has already been released but phrase it in a more easily understandable or usable way. For instance, this happens in the United States when Toxic Release Inventory data is released (Konar and Cohen 1997:114). The media publicize compilations of this data that reveal which companies have the highest emissions or do in-depth reports on specific firms to highlight problems of particular concern to their audiences. Because the media come in all shapes and sizes, they can tailor stories for any audience, from Wall Street investors to people living in the vicinity of a particular source of pollution. This tailored information can be of critical importance in educating the public and investors who, in turn, can pressure firms to innovate and reduce waste.

In countries where corporate environmental disclosure rules are either not in place or not systematically enforced, the media's job is considerably more difficult. In China, for example, journalists have been extraordinarily active in disclosing information about pollution in order to highlight its growing seriousness (Hong 1995:25).[28] They also assist the government, which has limited resources to reach a large population, in publicizing new environmental legislation and campaigns (Bo 1998:42). The resulting media reports, though no substitute for regular corporate disclosure mechanisms, educate the public, industry and policy makers and appear to influence policy-making in that country (Hong 1995:32; Bo 1998:41).

6. STRENGTHENING INFORMATION DISCLOSURE MECHANISMS

The disclosure rules embedded in regulatory, voluntary and market mechanisms are an important factor in recent improvements in corporate environmental performance and have also led to increased innovation and competitiveness in some firms. Nonetheless, as discussed above, weaknesses in these rules have limited their effectiveness on many occasions. As these weaknesses are only partially compensated for by the media and NGO efforts, it is worth asking how information disclosure rules might be strengthened to encourage higher levels of innovation and better environmental performance.

The key to effective disclosure rules and programs is a systemic approach which includes mechanisms for discovering environmental risks, for ensuring information reliability, and for publicizing and sharing information (Tietenberg 1997:10). It is only when these three mechanisms are in place that a solid foundation for action by the public, government, investors and firms exists. Though a wide variety of legal systems and philosophies exist, these mechanisms can be designed to function effectively at the local, national and regional level in any part of the world.

In the chemical industry, developing effective mechanisms for discovering environmental risks is probably the most difficult of the three. Some parts of the world, like the European Union, have regulations requiring specified information and risk assessments for all new chemicals as well as programs to test existing substances. However, even where such programs exist there is frustration at the considerable scientific uncertainty remaining about the effects of prolonged exposure to low concentrations of chemicals and at the slow progress being made in chemical testing (DETR 1998). In the United States, for instance, even basic toxicity testing results are still not available in the public record for almost 75 percent of the top-volume chemicals in use (EDF(d) 1997:7). Presently, efforts are underway to redesign risk assessment programs to make them more efficient and more accessible to the public.[29] The enormous amount of effort required to assess chemical risk necessitates regional and international cooperation and information exchange so that research efforts will not be duplicated.

The second part of a systemic approach to information disclosure is ensuring the reliability of information. While this is an issue even when information comes from government or university laboratories (thus necessitating well-designed peer review systems), it is arguably more important when the information comes from industry. In the case of chemicals, data on chemical use and emissions (as well as any other required by law) must be subject to third party review. The embedding of third party reviewed information disclosure in Responsible Care programs worldwide would be one effective way of promoting reliable disclosure in developing countries.

Finally, a systemic information disclosure strategy must include mechanisms for publicizing and sharing information. Embedded in these mechanisms must be programs that ensure information is understandable to the targetted audience. The needs of developing countries in this area deserve special mention. In terms of chemical risk assessment and reduction, it is vital that research done in industrialized countries be made available to developing country authorities and academics. While an increasing amount of chemical production is taking place in these countries, many have extremely limited trained staff and facilities for carrying out independent research. As well, information exchange systems for developing countries to share their experiences in areas like pesticide risk reduction urgently need to be developed.[30] The Co-ordinating Group for Chemicals Information Exchange is one body that is trying to develop mechanisms that will ensure effective information sharing (UNEP(b) 1998).[31] As for monitoring emissions during production, the major challenge in developing countries is ensuring that information is accessible and understandable for the public. Given the low rates of literacy in many of these countries, simple indicators are often most effective. The Indonesian government has been particularly imaginative in this area, colour-coding factories according to their emission levels. Low-emission factories get 'green' and 'gold' ratings, while highly polluting factories are labelled 'black' or 'red'. These ratings have been instrumental in bringing many 'black' rated factories into compliance with environmental regulations (Afsah, Laplante and Wheeler 1996).

One recently announced program that includes all three of these mechanisms is a joint initiative by the U.S. Environmental Protection Agency (EPA), the Chemical Manufacturers Association and the Environmental Defense Fund to test 2,800 chemicals for their environmental and health effects between 1999 and 2004. Under this program, chemical manufacturers are given 13 months to volunteer their products for testing. Any chemicals not volunteered during this period will be tested under EPA direction. The Environmental Defense Fund will be responsible for monitoring progress and will also disseminate information to the public over the Internet. This information will be detailed, outlining progress on a per chemical and per company basis (EDF(b) 1998). Incentives to share information internationally have been built into the program, as testing will not be required for those chemicals which are being tested outside of the United States and about which information is made available to U.S. companies (EDF(c) 1998).

A powerful tool for implementing systemic information disclosure programs is the Internet. In the area of chemicals, the relatively low cost information dissemination option that the Internet supplies makes it useful as a link between testing facilities, between various government regulators, and between NGOs and the public. The Internet could also be an important tool for providing chemicals training for agricultural extension workers and health professionals, particularly in developing countries. While the serious lack of access to telephones and computers in many developing countries is an obstacle at present, the importance of communications technologies for achieving other development goals is increasing their presence in schools and government offices. In short, if information disclosure programs are to have an impact in developing countries, a funding priority must be the provision of access to and training in interpreting and using chemical data. The Internet could greatly facilitate these activities.

7. CONCLUSION

In an increasingly complex corporate world, information disclosure is a powerful tool for promoting innovation, improving competitiveness and reducing waste. In the chemical industry, for instance, disclosure rules incorporated into regulatory, voluntary, and market mechanisms have contributed significantly to improvements in corporate performance. However, a focus on end-of-pipe information in many cases as well as problems such as information unreliability and inaccessibility have rendered many disclosure rules and programs only partially effective. The failure of information disclosure programs to reach small firms, both in industrialized and industrializing countries, also remains a problem. While this paper has focused on the failings of information disclosure programs in relation to chemical use and production, similar problems exist in the use and production of other goods and services.

Systemic disclosure programs that ensure the existence of risk assessment measures as well as the reliability and accessibility of information will be more effective than present programs in promoting innovation and enhanced environmental performance. Such programs are applicable in all sectors of the economy. However, in order for these programs to work in small and medium-sized firms, particularly in developing countries, technical and financial assistance will be necessary.

Ultimately, however, information disclosure rules in corporate governance mechanisms are not just about corporations. They are about the most basic issues of health, safety and environmental quality. The explosion in chemical production over the past fifty years has been linked to rising reproductive and developmental abnormalities in animal and human populations, and increasing rates of breast and other types of cancer. It is also deepening concerns over the general safety of the water and food supply. Yet, for all the laws and programs that exist to protect people and the environment around them from hazardous chemicals, we remain woefully ignorant of the nature and possible effects of the vast majority of chemicals. As Theo Colborn, co-author of the best-selling book on endocrine disrupting chemicals entitled Our Stolen Future recently remarked, "People would be shocked if they knew what was out there" (The Japan Times, 7 October 1998). It is time to develop systemic information disclosure systems that will shock politicians, investors and the public out of their complacency about the chemicals which surround us. It is only such systems that can begin to identify which chemicals are dangerous, how to protect ourselves from them and how to ensure the existence of a livable planet for our children.

ENDNOTES

1 The World Bank defines corporate governance as having to do "with ways in which various [stakeholders] in a corporation assure themselves that the corporation operates in accordance with stipulated goals and performance expectations. [Stakeholders] include owners, creditors, workers, consumers, government regulators, management and other groups who will benefit from the success of the corporation or who will be adversely affected by its inefficiency or misuse" (World Bank(a)).

2 Not surprisingly, high tech sectors like information technology tend to draw the most attention in this regard. For instance, in the United States, high tech has added almost a full point to economic growth over the past year (Mandel 1998:60). However, innovation is not limited to these sectors. The Dutch flower industry is a classic example of this fact. Despite its limited land area and unfavorable climate, the Netherlands is responsible for 65 percent of global flower exports. This is possible because Dutch growers "have innovated at every step in the value chain, creating technology and highly specialized inputs that enhance resource productivity and offset the country's natural disadvantages"(Porter and van der Linde 1995:130).

3 The shift away from command and control approaches to environmental problems has a number of facets. On one level it has resulted from a sense of frustration at the failure of these approaches to reduce environmental impacts. One report discussing command and control regulation in the European Union notes that "in spite of all the Directives and Regulations adopted by the EC and the international and national actions in this field, environmental quality is still deteriorating in the Community and worldwide" (www.emas.lu). The situation is particularly severe in many developing countries, where capacity to implement and monitor environmental regulation is limited. A further reason is the economy-wide trend toward deregulation spurred by both rapid technological change and trends in economic and political thought (Vogel 1996).

4 The increased use of disclosure requirements is particularly marked in the area of finance and banking. The New Zealand government has gone particularly far in this area, requiring all banks to release quarterly public disclosure statements. These must include credit rating information and are subject to external audit (Nicholl 1996). The Economist reports that U.S. banks are also to be subject to stricter disclosure rules in the near future. After the ignominious bailout of Long Term Capital Management (LTCM), regulators are poised to require U.S. banks to disclose more information regarding their exposure to high-risk hedge funds like LTCM.

5 For instance, the Massachusetts Toxic Use Reduction Act requires large quantity users of toxic chemicals to create an inventory of the chemical flow in and out of production processes and, at the same time, develop plans to reduce that

flow for each of these processes. Its objective is that "through the process of planning, firms will identify more efficient production methods that will both prevent pollution and save money" (Roy 1993). In this way, the Act aims "to sustain, safeguard and promote the competitive advantage of Massachusetts businesses, large and small" (TURA, M.G.L. c.211).

6 For instance, under Japan' s Basic Environmental Law, corporations are specifically encouraged to develop voluntary action plans (Imura(b)1998:9), thus blurring the line between regulatory and voluntary measures. In the Netherlands, pollution abatement targets set out in voluntary agreements are eventually incorporated into emission licenses (Borkey and Leveque 1998:14).

7 While voluntary mechanisms are non-regulatory, in most situations they are undertaken either to avoid regulation or with the understanding that a failure to abide by them will lead to the imposition of regulation.

8 National level freedom of information laws exist in such countries as Australia, Canada, Ireland and the United States. The Republic of South Africa is presently considering introducing such a law, and campaigns in Great Britain and Israel are underway to bring about such legislation. For further information, see www.missouri.edu/~foiwww/intl.html. The European Union has a Directive specifically on freedom of access to information on the environment.

9 See Deirdre H. Robbins, "Doing Business in the Sunshine: Public Access to Environmental Information in the United States", Review of European, Comparative and International Environmental Law vol.3 no.1(1994), 26-35 for a discussion of trade secrets and the U.S. Emergency Planning and Community Right to Know Act (EPCRA).

10 PRTRs are not necessarily regulatory measures, but in practice usually take a regulatory form. They presently exist in a at least six countries: the Australia, Canada, England, France, South Korea, and the United States. Mexico has a partial PRTR which covers 140 chemicals. The information provided by companies under Mexico's program will not be made available to the public for 3 years due to the alleged "lack of government resources to administer this program and the lack of public demand for toxics data" (Sissell 1998). In addition to these national level programs, at least one local government, Japan's Kanagawa Prefecture, has enacted a PRTR.

11 A number of U.S. environmental groups provide PRTR information in a readily usable form for concerned citizens to use. Deserving special mention is the Environmental Defense Fund's Chemical Scorecard. Among other things, it allows users to enter their zip codes and see which facilities are the biggest polluters in their areas. In addition, it provides health information on all of the Toxic Release Inventory chemicals (approximately 650 of them) plus over 4000 others. For further information, see www.scorecard.org. Friends of the Earth has a similar Internet site covering England and Wales. See www.foe.co.uk/cri/html/postcode.html.)

12 The PIC Convention is based on the voluntary PIC procedure which was accepted by all Food and Agriculture Organization (FAO) and United Nations Environment Program (UNEP) member countries in 1989. Because the Convention has not yet gone into effect, trade in banned or severely restricted chemicals is still monitored by the voluntary PIC agreement. For a short introduction to this agreement, see "Prior Informed Consent: A Brief Overview of What It Is and How it Operates" at www.fao.org/ag/agp/agpp/pestici/pic/download/brief.doc)

13 The Pesticide Action Network notes, in particular, the difficulties faced in developing countries in this regard: "To be considered for inclusion as a severely hazardous pesticide formulation, the agreement requires that information must be compiled on specific incidents in developing countries, the adverse effects, and the way in which the formulation was used. NGO experience confirms that evidence at such a detailed level is extremely difficult to compile. Incidents take place far from medical facilities and regulatory authorities, and many farmers are unaware of what pesticides they are using since pesticides are often sold in unmarked sachets. In many African countries, poison control centers have not been established and collecting information on incidents is virtually impossible" (PANUPS 1998).

14 The importance of informed agricultural and health workers should not be underestimated. Each year, there are an estimated 2 million cases of pesticide poisoning in the Asia Pacific region. 40,000 of these cases end in death. While a number of factors are responsible for these factors, "ignorance of risks" is key (Pearce and Warford 1993). As high rates of illiteracy will likely persist in many rural areas, trained agricultural and health workers can play a vital role in disseminating information about chemical risk.

15 Different authors use different terms for the various voluntary mechanisms. For instance, Labatt and Maclaren use the terms self-regulation, voluntary challenge, and voluntary agreement to refer to unilateral commitments, public voluntary schemes and negotiated agreements, respectively (Labatt and Maclaren 1998:192-3).

16 The Investor Responsibility Research Center notes that the oil company Exxon has 69 percent of its assets outside of the United States. The corresponding figure for Dow Chemical is 57. These firms, however, are exceptional in that they report extensively on the environmental performance of their non-U.S. operations (Cassidy).

17 The most well known of these guidelines are those produced by the International Chamber of Commerce in 1990, the Guidelines for Environmental Reporting put out by the Public Environmental Reporting Initiative (PERI) in 1993, and the Coalition for Environmentally Responsible Economies (CERES) principles of 1992. Of these, the CERES principles require signatories to provide the most specific and extensive information (Davis-Walling and Batterman:868).

18 Many industries have either directly copied or at least incorporated some aspects of Responsible Care in their environmental management programs. In the former group is the Canadian Association of Chemical Distributors, in the latter the American Petroleum Institute (Labatt and Maclaren 1998: 194-5)

19 The Canadian Chemical Producers Association plans to reduced total emissions two thirds from 1992 levels by 2001.
www.ccpa.ca/reports/1997/nerm/english/default.html)

20 This compares to the Canadian Chemical Producers Association, which can point to concrete, industry-wide reductions. Between 1992-1996, emissions were cut in half, despite an almost 20 percent increase in chemical production over the same period.

21 The U.S. Public Interest Research Group (PIRG), for instance, argues that most chemical companies in the country have not yet adopted the principles of Responsible Care and that many of them continue to oppose stronger freedom of information legislation that would provide local communities with more complete information on the toxic chemicals produced and released by industry (PIRG:1-2).

22 The lack of anti-trust legislation can have particularly serious implications for small and medium enterprises. In Austria, voluntary environmental agreements among large companies in the waste sector led to distorted prices and competitiveness problems for smaller companies (European Environment Agency 1997).

23 No doubt for this reason, ISO 14001 is attracting attention in developing countries as well. A number of major companies, both domestic and foreign-owned, in Mexico and in ASEAN countries have already received ISO 14001 certification (World Bank 1998; Wong 1998).

24 The Palestinian Ministry of Economy and Trade suggests that ISO certification for that country's firms can be achieved only if there is a coordinated, economy-wide strategy in place. It argues that agencies in the public sector will have to institutionalize ISO management and auditing standards and then transfer them into the private sector in a well planned fashion. This transfer would be accompanied by the provision of information and some technical assistance. The Ministry argues that this will help local firms develop quality products for the domestic market in the short term and qualify for ISO certification in the long term (Palestine Economic Pulse 1998). This type of approach may be applicable in other developing countries as well.

25 Negotiated agreements between industry and NGOs only, without government involvement, is rare in the chemical industry but occurs more frequently in industries like forestry and fisheries. For instance, WWF and other environmental NGOs have joined up with industry to set up a Forest Stewardship Council and a Marine Stewardship Council.

[26] Imura classifies these agreements as public voluntary schemes. However, though local governments initiate these schemes, quantitative emissions levels and pollution control measures are arrived at through negotiation. In this sense, they fall more comfortably in a negotiated agreement category.

[27] For example, 97 percent of the firms in the Canadian Chemical Producers Association have ARET membership.

[28] The Chinese press also compensates for a lack of adequate dispute settlement mechanisms. Parties that feel their cases have not been adequately heard press for the publication of their cases in order to raise public sympathy and put pressure on the relevant administrative bodies (Hong: 28).

[29] One of these is the work of the U.S. Environmental Protection Agency's Endocrine Disrupter Screening and Testing Advisory Committee. Its final report recommends that chemicals be tested in several stages: first an automated test, followed by other tests only if the first test shows "a hint of danger". The tiered testing will make the process more efficient, but the task will be far from simple. The Committee recommends that 15,000 chemicals be tested (The Japan Times, 7 October 1998).

[30] Developing countries appear to have much to learn from each other, particularly in terms of pesticide risk reduction. Many of these countries face similar problems in designing and implementing programs in this area. These include difficulties in creating training materials appropriate for subsistence farmers who are unable to read and a serious shortage of extension personnel (OECD(c) 1996:17). Solutions to these problems developed in one country may be applicable in others and must be shared.

[31] The Co-ordinating Group for Chemical Information Exchange falls under the Inter-Organization Programme for the Sound Management of Chemicals (IOMC). The IOMC was established in 1995 by UNEP, ILO, FAO, WHO, UNIDO and the OECD to improve information exchange on chemicals (OECD(c) 1996).

REFERENCES

Afsah, Shakeb, Benoit Laplante and David Wheeler. "Controlling Industrial Pollution: A New Paradigm". October 1996. (www.worldbank.org/NIPR/work_paper/1672/wp1672.pdf)

Afsah, Shakeb, Benoit Laplante and Nabiel Makarim. "Program-Based Pollution Control Management: The Indonesian PROKASIH Program". December 1995. (www.worldbank.org)

Benneh, George, William B. Morgan and Juha I. Uitto, eds. *Sustaining the Future: Economics, Social and Environmental Change in Sub-Saharan Africa*. Tokyo: United Nations University Press, 1996.

Birnie, Patricia W. and Alan E. Boyle. *International Law and the Environment.* Oxford: Clarendon Press, 1992.

BMU (German Environment Ministry). "Environmental Protection and Financial Services: Positions and Activities of the Federal Environment Ministry". (www.bmu.de/englisch/index.htm)

Bo, Wen. "Greening the Chinese Media", *China Environment Series* (Woodrow Wilson Center Environmental Change and Security Project), Issue 2 (Summer 1998), 39-44.

Borkey, Peter and Francois Leveque. *Voluntary Approaches for Environmental Protection in the European Union*. Workshop on the Use of Voluntary Approaches in Environmental Policy (OECD). March 1998.

Boyd, James. "Searching for the Profit in Pollution Prevention: Case Studies in the Corporate Evaluation of Environmental Opportunities". Resources for the Future, 1998. (www.epa.gov/opptintr/acctg/paper2b.pdf)

Bruner, Laura J. and Michael J. Burns. "The ISO 14000 Series: Business-Friendly Environmentalism", *Environmental Regulation and Permitting*, Spring 1998, 17-19.

Burnett, Miles. L. "The Pollution Prevention Act of 1990: A Policy Whose Time Has Come or Symbolic Legislation?", *Environmental Management* vol.22, no.2 (1998), 213-224.

Campanale, Mark A. "Green Investment: Incentives for Disclosure", *Review of European Comparative and International Environmental Law*, vol.3 no.1 (1994), 43-48.

Cassidy, Alison. "Disclosure of Environmental Information about S&P 500 Companies' Non-U.S. Operations". (www.irrc.org/eisweb/eismain.htm)

Chaytor, Beatrice and Mathias Wolkewitz. "Participation and Priorities: An Assessment of Developing Countries Concerns in the Trade/Environment Interface", *Review of European Comparative and International Environmental Law*, vol.6 no.2 1997, 157-162.

Crossley, Rachel and Jonathan Points. *Investing in Tomorrow's Forests*. 1998. (www.panda.org/forests4life/news/report.pdf)

Dasgupta, Susmita, Benoit Laplante and Nlandu Mamingi, "Pollution and Capital Markets in the Developing World", The World Bank Development Research Group, 1997.
(www.worldbank.org/nipr/work_paper/market/markets.pdf)

Davis-Walling, Paige and Stuart A. Batterman. "Environmental Reporting by the Fortune 50 Firms", *Environmental Management* vol.21 no.6 (1997), 865-875.

Davies, Terry. "Critically Evaluating America's Pollution Control System". (www.rff.org/resources_articles/files/regpoll2.htm)

Day, Robert M. "Beyond Eco-Efficiency: Sustainability as a Driver for Innovation". *Sustainable Enterprise Perspectives*. March 1998. (www.wri.org/meb/sei/beyond.html)

DETR (Department of the Environment, Transport and the Regions, U.K.). "Sustainable Production and Use of Chemicals: Consultation Paper on Chemicals in the Environment", April 1998. (www.environment.detr.gov.uk/sustainable/chemicals/consult/03.htm#2)

Dupont. "Safety, Health and the Environment: 1997 Progress Data". (www.dupont.com/corp/environment/data.html#regp)

EDF (Environmental Defense Fund) (a). "Toxic Release Inventory Shows Need for Pollution Prevention". *EDF News Release*, 18 June 1998. (www.edf.org/pubs/NewsReleases/1998/Jun/d_toxics.html)

EDF (b). "Industry to Test 2,800 Major Chemicals for Health, Environmental Effects". *EDF News Release*, 9 October 1998. (www.edf.org/pubs/NewsReleases/1998/Oct/b_cma.html)

EDF (c). "Additional Details on the Voluntary Testing Program". October 1998. (www.edf.org/pubs/FactSheets/1_cma.html)

EDF (d). Toxic Ignorance. 1997. (www.edf.org/pubs/reports/toxicignorance) "EMAS: An Overview, Progress-to-Date and Current Issues". 1998. (www.emas.lu)

Environment Canada. "1996 NPRI Summary Report Executive Summary". (www2.ec.gc.ca/pdb/npri/96)

EPA. "EPA's Toxic Release Inventory". (www.epa.gov/reg3artd/tchem/tri.htm)

European Environment Agency. "Environmental Agreements, Environmental Effectiveness". *Environmental Issues Series* no.3, vol.1. Copenhagen 1997. (www.eea.dk/frdocu.htm)

Greenpeace International. "The Decline of Corporate Accountability". (www.greenpeace.org)

Hammond, Allen. *Which World? Scenarios for the 21st Century. Global Destinies, Regional Choices*. Washington D.C.: Island Press, 1998.

Hong, Mei. "Legal Gateways for Environmental Protection in China", *Review of European, Comparative and International Environmental Law* vol.4 no.1 (1995), 22-32.

ICCA (International Council of Chemical Associations). "Industrial Perspectives on International Chemicals Management". (www.icca-chem.org/issues.htm)

Imura, Hidefumi (a). "The Use of Voluntary Approaches in Japan: An Initial Survey", presented at the Workshop on the Use of Voluntary Approaches in Environmental Policy (OECD), 1-2 July 1998.

Imura, Hidefumi (b). "The Use of Unilateral Agreements in Japan: Voluntary Action Plans of Industries Against Global Warming", presented at the Workshop on the Use of Voluntary Approaches in Environmental Policy (OECD), 1-2 July 1998.

IRRC (Investor Responsibility Research Center". "Green Investing: Good Environmental Sense Makes Good Financial Sense". (www.irrc.org/eisweb/eismain.htm)

Konar, Shameek and Mark A. Cohen. "Information as Regulation; the Effect of Community Right to Know Laws on Toxic Emissions", *Journal of Environmental Economics and Management* vol. 32 (1997), 109-124.

Labatt, S. and V.W. Maclaren. "Voluntary Corporate Environmental Initiatives: A Typology and Preliminary Investigation", *Environment and Planning C: Government and Policy*, vol 16 (1998), 191-209.

Mandel, Michael J. "You Ain't Seen Nothing Yet", Business Week, 31 August 1998, 60-63.

Mazurek, Janice. "The Use of Unilateral Agreements in the United States: The Responsible Care Initiative" presented at the Workshop on the Use of Voluntary Approaches in Environmental Policy (OECD), 1-2 July 1998.

Morrow, Hugh. "Reducing Cadmium Risk in the U.S. through and Industry Sponsored Nickel/Cadmium Battery Collection and Recycling Program" in Proceedings of the OECD Workshop on Non-Regulatory Initiatives for Chemical Risk Management, 10-12 September 1996, 27-34. (www.oecd.org/ehs/ehsmono/06E71417.pdf)

Murty, M.N. "Environmental Regulation in the Developing World: The Case of India". *Review of European, Comparative and International Environmental Law*, vol.4 no.3 (1995), 330-337.

Nicholl, Peter. "Market Based Regulation of Banks: A New System of Disclosure and Incentives in New Zealand". *Public Policy for the Private Sector*, Note no.94, October 1996. (www.worldbank.org/html/fpd/notes/94/94nichol.pdf)

OECD(a). *Pollution Prevention and Control Extended Producer Responsibility in the OECD Area Phase 1 Report: Legal and Administrative Approaches in Member Countries and Policy Options for EPR Programmes*. (OCDE/GD(96)48). 1996.

OECD(b). *Pollution Release and Transfer Registers (PRTRs): A Tool for Environmental Policy and Sustainable Development: Guidance Manual for Governments*. (www.oecd.org//env/docs/en/gd9632.pdf)

OECD(c). *Activities to Reduce Pesticide Risks in OECD and Selected FAO Countries. Part I: Summary Report*. OECD Environmental Health and Safety Publications, No.4, 1996.
(www.oecd.org/ehsmono/07e69625.pdf)

Palestine Economic Pulse. "Private Sector Update: ISO- International Standards in Palestine", 1998. (www.palecon.org/pulsedir/march/private.html)

PANUPS (Pesticide Action Network). "NGOs Call for Rapid Implementation of Prior Informed Consent", 14 September 1998. (www.dejanews.com)

Pearce, David W. and Jeremy J. Warford. "Economics, Environment and Sustainable Development: A Summary". 1993.
(www.worldbank.org/html/extpb/reports/WorldEnd2.html)

Porter, Michael E. and Claas van der Linde. "Green and Competitive: Ending the Stalemate", *Harvard Business Review*, September-October 1995, 120-134.

PIRG(a). "Trust Us, Don't Track Us: Executive Summary". (www.pirg.org/enviro/toxics/track98/page1.htm)

PIRG(b). "Trust Us, Don't Track Us: Materials Accounting Data: Successful State Programs". (www.pirg.org/enviro/toxics/track98/page9.htm)

PIRG(c). "Right to Know Data Show Toxic Wastes Increasing, Industrial Pollution Prevention Efforts Failing". 18 June 1998. (www.pirg.org/enviro/toxics/rtk/epatri98.htm)

Ranganathan, Janet. "Sustainability Rulers: Measuring Corporate Environmental and Social Performance". *Sustainable Enterprise Perspectives*. May 1998. (www.wri.org/meb/pdf/janet2.pdf)

Riordan, James. "Accelerated Reduction/Elimination of Toxics (ARET): Government Perspective" in Proceedings of the OECD Workshop on Non-Regulatory Initiatives for Chemical Risk Management, 10-12 September 1996, 45-48. (www.oecd.org/ehs/ehsmono/06E71417.pdf)

Robbins, Deirdre H. "Doing Business in the Sunshine: Public Access to Environmental Information in the United States". *Review of European, Comparative and International Environmental Law*, vol.3 no.1, 26-35.

Roy, Manik. "Toxics Use Reduction: Act Overview", 1993. (www.magnet.state.ma.us/dep/bwp/dhm/tura/turaover.htm)

Sissell, Kara. "Mexico Starts Up Program". *Chemical Week* vol.160 no.20, 27 May 1998.

Smart, Bruce. *Beyond Compliance: A New Industry View of the Environment*. World Resources Institute, 1992.

The Economist(a). "Long-term sickness?", 3 October 1998, 91-93.

The Economist(b). "The Burning Issue", 25 July 1998, 68-69.

The Japan Times. "Aiming for Good Corporate Governance", 6 August 1998.

The Japan Times. "Testing of 15,000 Chemicals Recommended",7 October 1998.

Thomas, Jim and Dick Constable. "ISO 14000: A New Approach to International Environmental Standards". *Environmental Regulation and Permitting*, Spring 1996, 3-11.

Tietenberg, Tom. "Information Strategies for Pollution Control"presented at the Eighth Annual Conference of the European Association of Environmental and Resource Economists", Tilburg University, The Netherlands, 26-28 June 1997. (www.colby.edu/personal/thtieten/info_strat_full.pdf)

UNEP (a). *UNEP Statement by Financial Institutions on the Environment and Sustainable Development (As Revised- May 1997)*. www.unep.ch/eteu/envr-fin.html

UNEP (b). "Information Exchange and Capacity Building". 1998. (irptc.unep.ch/irptc/docs/news/t22.html)

USIA. "U.S. Will Sign Rotterdam Convention", 11 Sept, 1998. (www.dejanews.com)

Vogel, Steven K. *Freer Markets, More Rules: Regulatory Reform in Advanced Industrial Countries*. Ithaca: Cornell University Press, 1996.

Wong, Eeling. "The Environmental Dimension of Corporate Finance". (www.memegroup.co.uk/magazine/09_september/biweek_1/enviro_riet_finance.html)

World Bank (a). "Corporate Governance". (http://www.worldbank.org/html/fpd/privatesector/corpgov.htm)

World Bank(b). "ISO 14000 Standards for Environmental Management". *New Ideas in Pollution Regulation*: Appendix 2. 23 September 1998. (www.worldbank.org/NIPR/work_paper/1877/WP1877web-11.htm)

West, Steve. "Accelerated Reduction/Elimination of Toxics (ARET) (Industry Perspective)" in Proceedings of the OECD Workshop on Non-Regulatory Initiatives for Chemical Risk Management, 10-12 September 1996, 49-51. (www.oecd.org/ehs/ehsmono/06E71417.pdf)

Chapter 6

SUSTAINABLE WATER STRATEGIES FOR NEW DEVELOPMENT PATTERNS

MAKOTO MURASE
International Rainwater Catchment Systems Association

1. INTRODUCTION

1.1 The Problem

Communities throughout the world face problems related to water. The problems vary only in degree and include such concerns as droughts and floods, drying up of springs and groundwater, eutrophication of lakes and ponds, pollution of rivers, contamination of groundwater, and unsafe drinking water.

Water is a key to the sustainable development of society. According to the United Nations, if current trends continue the human population will reach over nine billion people by the middle of the twenty-first century. Most of the increase is now taking place in developing countries, and it is believed that this trend will continue. Along with the population increase an enormous growth in demand for food will arise, and corresponding to that, huge demand for water. Another ongoing trend is the accelerating concentration of people in cities; it is thought that by about 2025 close to two-thirds of the world population may be living in cities. It is no exaggeration that new urban development patterns and sustainable water strategies will both be important keys to global environmental strategies. Although this paper focuses on water in new urban development patterns as its main topic, the discussion is equally applicable to new patterns of rural development. In particular the issue of cycle capacity is important for agriculture.

To begin the discussion, consider the following examples of the sequence of problems as they typically emerge relating to urban development and water resources.

K. Matsushita (ed.), Environment in the 21st Century and New Development Patterns, 147–162.

1.1.1 Fate of Large Dam Developments

A city lacks sufficient water. It seeks water from the upper reaches of rivers. A large dam is constructed. Urban concentration of the population accelerates. A greater water shortage occurs. The dam is insufficient to fill needs. A more remote area upstream must be developed to supply water. Villages will disappear beneath the waters. Voices are raised in protest from those areas. A very long period elapses before the new dam can be completed. The construction costs of the dam soar. The dams that have already been constructed begin to fill with silt and their effective water storage capacity shrinks. Do we really need more large dams?

1.1.2 Water Treatment Systems and Rivers that Can't Handle Rain

It is desirable to eliminate urban rainwater quickly. Good sewer systems are constructed. Urbanization advances further. During heavy rains the sewers back up. Sewer water overflows in the city. New construction sends the rainwater out into the rivers more quickly. The rivers flood. Winding rivers must be straightened to improve flow. Rivers must be widened and dug deeper. Even so, the rivers flood. Can floods really be prevented just by improving the sewer systems and altering the rivers?

1.1.3 Fate of Sewer Water Treatment

Urban rivers become polluted. More effective water treatment becomes desirable. It makes sense to have water treatment for the entire watershed. Water stops flowing into rivers upstream. Downstream most of the water flowing in rivers has been treated already. The water looks clearer but it is far from clean. More extensive water treatment is needed That requires major inputs of energy and finances. The new treatment system produces a large amount of sewage sludge. There is no place to dispose of the sludge. Is this the fate of large water treatment systems?

1.1.4 Limits of Water Supply Management

Water supplied tastes bad. Treat it with activated carbon. It still tastes bad. A decision is made to treat water with ozone. Enormous costs are incurred in constructing a treatment plant. Maintenance and operating costs pile up. Water rates must be increased. In the end people give up and purchase bottled water.

These examples of the progression of water problems present only a few possible scenarios. The responses to the problems presented are not necessarily the inevitable outcomes. However, these scenarios indicate that the mere reliance on technological fixes to address each new problem that arises only result in postponing fundamental solutions to the problems. The true nature of urban water problems lies in water resource policies that throw away the rainwater that falls on a city, make the city dependent on upstream sources in watersheds, and ignore the hydrological cycle approach to water resource management. If policies do not change fundamentally to break the vicious cycles, there is no way forward to sustainable new urban development patterns.

2. A CHANGE IN THINKING

2.1 Shifting from Water Dependence to Self-Reliance

2.1.1 From Off-Site to On-Site

Until the present day the water resource policies of many cities of the world have been based on the idea of drawing urban water from great distances of up to hundreds of kilometers away. Tokyo is no exception. Since it was accepted that the city's water supply was inadequate, dams far from the city became necessary. Most of Tokyo's piped water supply comes from dams along the Tone (pronounced "toe-nay") River system, about 150 kilometers from the center of the city. According to the 1999 water budget drawn up by the metropolitan government water resources that would be an equivalent of 828 millimeters of rainwater were brought to Tokyo from outside the city limits. This is equivalent to 59 percent of Tokyo's annual average rainfall of 1405 millimeters, and most of it came from the Tone River water system.

But this policy of increasing dependence on the upper reaches of the watershed is beginning to lead to a dead end. In 1987 rain did not fall upstream for a long time, causing a serious water shortage in Tokyo and surrounding the Kanto area. Saying that development of the upper reaches of the Tone River watershed had reached its limits, Shunichi Suzuki, then governor of Tokyo, proposed the Shinano River Watershed Scheme. The main concept was to draw water from the Shinano River, which is outside the Tone watershed, into the Tone River. The then governor of Niigata Prefecture, Takeo Kimi vehemently opposed this scheme, saying that "seeking water resources in another prefecture just because your own water is in short supply is too simplistic." Dam construction in the upper watershed would submerge many houses and fields and mountain woods, and destroy the culture of hill country villages that existed over centuries. Because it is never easy to gain approval from people whose homes will be submerged, completion of the dam construction would take a long time. Compensation to these people, and compensation for the submerged land increases over time, resulting in higher construction costs.

There are no further prospects for the development of large dams in the upper reaches of the Tone River, and construction of dams already proposed is only progressing slowly. In addition, dams that have already been constructed are filling with silt, reducing their water storage capacity. Tokyo must change its approach and reduce its dependence on water resources in the upper watershed. New water policies should aim for water self-reliance by using nearby sources as much as possible, including rainwater, groundwater and recycled sewer water, and progress with water conservation is essential.

Seeing rainwater as only a nuisance, Tokyo had always sought ways to eliminate it quickly through the sewers. This approach virtually ignored the potential to use rainwater as a water resource, although the 2.5 billion cubic meters of rain that fall on Tokyo in a year exceed the 2 billion cubic meters of piped water consumed. This rainwater should be taken as a resource and actively used. The lesson that we need to claim urban rainwater as a resource and to change from "off site" to "on site" water sourcing is the lesson from the experiences of Tokyo's past water resource policies - for Japan and for the world .

2.1.2 From Life Line to Life Point

When a city has a high degree of dependence on distant water sources, and then there is a long period without rainfall in the watershed, the normal functioning of the city is quickly paralyzed. Reliance on a water pipeline to the city also makes it vulnerable. A city totally reliant on a pipeline drawing water from a distant water source is also extremely vulnerable in the face of a large-scale natural disaster. Japan's catastrophic Hanshin-Awaji earthquake in 1995 showed how this can happen in reality. The city of Kobe, which depends on a pipeline from the distant Biwa Lake for most of its water, was severely affected when the earthquake destroyed the supply line. During the approximately one month it took to repair the pipeline, Kobe's municipal functions were completely paralyzed. However, while water from this pipeline was cut off well water from within the city served as an alternative water source. For example, in Kobe's city hall, where the Disaster Response Center was based, the toilets continued to function. Why? Because the building had been designed with a dual water supply system using both pipe water and well water; the toilets used well water. The huge earthquake had not damaged the well water system, and the toilets continued to operate.

A transition is needed from the "life line" to the "life point" approach. This is the lesson from the Kobe disaster. Numerous small-scale and decentralized water sources should be secured as "life points" within the city, drawing on rainwater, groundwater and other sources. This approach will create cities that are strong in the face of drought and earthquakes.

2.2 From Interruption of Local Hydrological cycles to Coexistence

2.2.1 Changes in Urban Structure

Many of the world's large cities now face problems with urban flooding as a result of the accelerating pace of urbanization. Often the core of this problem is the destruction of the local hydrological cycle, which consists of rain falling and gradually seeping into the ground to become groundwater. A rich supply of groundwater produces springs and underground rivers. The figure based on records for the wards of Tokyo between 1968 and 1991, shows the ratio of land area where rainwater does not permeate compared with the area where it does. In 1991 the ratio of land area impermeable to rainwater reached 81.8 percent. An increase in

Figure 6-1. Impermeable Land Area and Rainwater Permeation in Tokyo

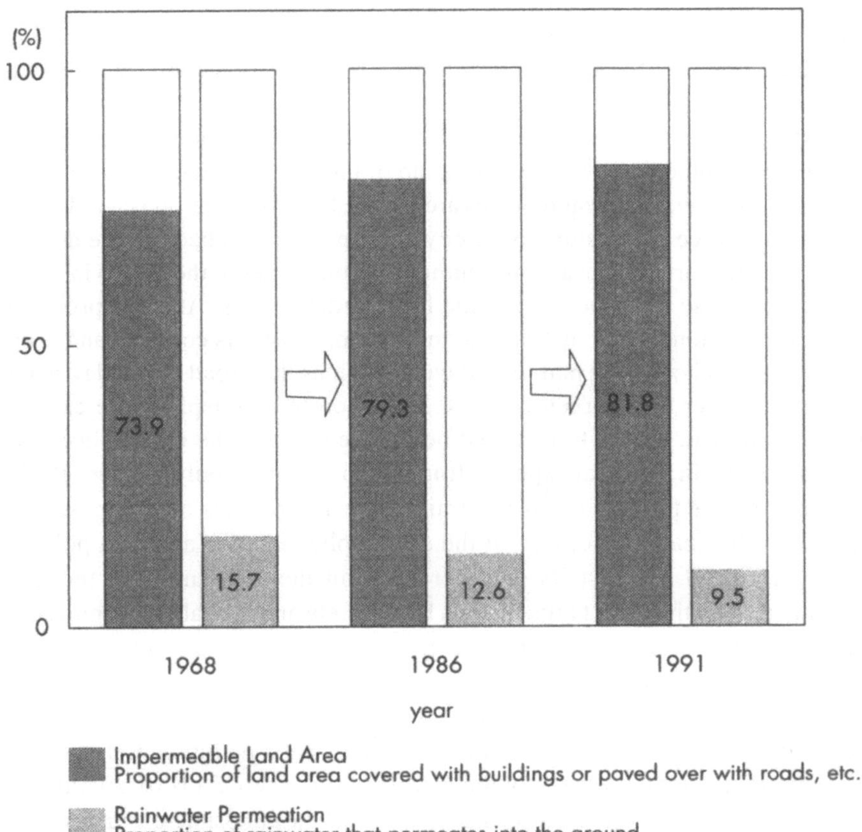

Source: Tokyo Metropolitan Environment Protection Bureau, Tokyo Metropolitan Water Environment Protection Plan, June 1998.

this ratio speeds up the rate of rainwater flow, resulting in its rapid concentration in storm drains and rivers. This increases the frequency of sewers backing up and flooding on small and medium rivers whenever heavy rain falls over a concentrated period of time. Even if flooding does not occur, the heavy rain causes an outpouring into the rivers and streams, from countless sewer outlets and pump houses, causing degradation of the urban river water. In Tokyo, New York and other big cities, prevention of river water pollution has become an important problem.

The design of cities tends to interrupt the local hydrological cycle and reduce the amount of rainwater permeating underground. Of the 1,405 millimeters of rainwater falling on Tokyo annually, only about 360 penetrate underground. Excess drawing of water from Tokyo's deep water table has already led to a serious problem with land subsidence, and the drying up of many springs and shallow wells. In response, Tokyo regulated the use of groundwater, an act which solved the problem of subsidence. However, the springs remained dry and the declining trend in rainwater penetration has continued, though at a reduced speed. In fact, the amount of groundwater that flows into rivers and ponds from springs today is thought to be no more than 30-100 millimeters in the wards of Tokyo. Because many springs in the city have dried up, the natural result is that urban rivers receive much less water than before from springs.

Construction that destroys the local hydrological cycle has various effects on the urban environment. One is the drying up of the city. This happens because the springs dry up, rivers and watercourses are covered, and vegetation is cut down. As an important source of moisture in the city atmosphere is choked off, the drying of the city escalates further. The average humidity at the center of the Tokyo in January has fallen by close to 20 percent in the last hundred years. Another problem is urban heat pollution. Being in Tokyo in mid-summer with its concrete and asphalt is like being on a hot frying pan. In hot weather an asphalt road at midday reaches temperatures over 60 degrees Celsius. As concrete and asphalt are excellent conductors of heat, heat is transmitted below the surface; the ground becomes a giant heat reservoir. The heat expelled from air conditioners only worsens the heat problem. Now the number of 'tropical nights' in Tokyo rises on a one way road along with urbanization. Because of the drying phenomenon and heat pollution, downtown Tokyo has lost its summer evening dew. Changes in the urban environment, with the earth tamped down by concrete and asphalt, have profoundly affected the city's ecological balance. As just one example of impacts, the number of swallows in Tokyo has plummeted.

The destruction of local hydrological cycles brought on by urban development is the basic cause of many problems: urban flooding, water pollution in rivers caused by the overflow from sewers when it rains, the drying up of springs, the transformation of the city into a heat island, drying of the atmosphere and destruction of the city's ecological environment. A comprehensive solution to this problem will require a radical change in urban design. Urban water problems must be revisited from the perspective of the local hydrological cycle, and fundamental solutions will emerge through the rebirth of these local hydrological cycles. Typical urban planning in the past meant more sewer facilities and river 'improvements' based on the assumption that more rainwater should be drained away. In contrast, from now on, from the stand point of preserving local hydrological cycles, it will be better to find ways to retain rainwater and to actively promote its permeation into the ground by limiting increases in the ratio of rainwater drained away, such as through preserving the soil and vegetation.

2.2.2 Stop Underground Development

In working out a change in policy, to move from the interruption of the local hydrological cycles to coexistence, a change is needed in the design of cities. Accordingly, regulations relating to underground development must be strengthened. If after the rainwater has managed to seep into the ground, large quantities are blocked by underground structures, the revitalization of the local hydrological cycle in the future will be difficult. This is especially true of large cities in the monsoon regions of Asia which are built on river deltas rich in groundwater. Tokyo is no exception. At present, every day large quantities of groundwater in Tokyo wards seep into subways and sewer drains, and into underground roads and buildings. As underground structures are placed directly in the shallow groundwater table, the groundwater seeps through joints and cracks in the structures. Large quantities of groundwater seep in, especially in underground structures that cut under riverbeds. It is estimated that about 900,000 cubic meters per day of groundwater seep into underground structures in the ward area of Tokyo, equal to about 26% percent of the 3.5 million cubic meters of piped water consumed each day in the ward area of Tokyo in 1997. With the aging of underground structures, this amount is expected to increase in the future. As this water is pumped out, over time new artificial groundwater courses will be formed, raising concern about great impacts on groundwater flows.

Developing countries at present are suffering from serious paralysis in transportation due to rapid increases in the number of vehicles on roads. Research is being done to introduce subway train lines to solve this problem. Also attempts are being made to promote large-scale drainage facilities as a way to prevent pollution of river water. However, based on the lessons of Tokyo's experience, from the point of view of preserving the local hydrological cycles, it would be better to consider trains running on the surface rather than underground. Instead of burying large sewer mains deep underground, it would be better to provide small-scale common treatment facilities which would not require massive construction, as these so-called individual sewers would not require large sewer mains. If for some reason underground sewers are necessary, it would be better if they were small in scale and not buried deep underground.

The problems of underground structures collecting groundwater don't end there. Structures built underground so far have been mainly in the shallow groundwater table between 10 and 20 meters deep. However advances in engineering technology are making it possible to build structures deeper; this means reaching into deeper groundwater reserves. This deeper level is now being targeted for development. Before the collapse of Japan's 'bubble' economy in the early 1990s which had overheated in response to the sudden rise in land prices, deep underground developments were frequently discussed. On this topic, the Environment Agency of Japan, in the interim report of a special committee to study the issue, stated that the 'strata that is being considered for deep underground development in large cities is largely the prime groundwater retention zone. If development is carried out in a disorderly way, there is reason for concern that it may cause ground subsidence and hamper the utilization of groundwater; water pressure of water at deep levels is likely to increase, and the construction of underground structures may cause damage to the subsurface flow of water.' The report indicated that construction in deep groundwater retention zones may result in water seeping into the underground structures over a long period of time, and have serious impacts on subsurface water courses. As groundwater at deep levels has seeped down over long periods of time, it is of very good quality and to this day is used as a water source in cities in monsoon deltas throughout the world. Seen from the point of view of conserving the quality and quantity of this water, development of the so-called 'geofront' (in contrast to the waterfront), or deep underground development, should be strictly controlled.

2.2.3 A Change in Drainage Policy

In order to revive local hydrological cycles, current drainage policies must be fundamentally reconsidered. First, a shift is needed from the present large-scale to small-scale drainage sewers, in particular systems based at the household level. In the past, drainage in large cities has been based on collecting water from a large area and processing it in large-scale treatment centers. Certainly this may be efficient in a very high-density city like Tokyo, but Tokyo is a very special case. In many situations this type of system may be inefficient and incur high costs. The large-scale drainage systems promoted by Japan provide classical examples: a giant treatment center is built downstream, and an extensive network of underground drainage sewers extends upstream for many tens of kilometers. Construction of this type of system requires very long timeframes to complete and incurs enormous costs. In contrast, individual drainage systems based on small-scale common treatment tanks have short construction periods and can be completed at a fraction of the cost. In the case of large-scale drainage sewers, because the polluted water is taken from a neighborhood and transported to a distant treatment center, it is very difficult for the drain user to have any awareness of the polluted water and where it is going, or to sense the effect of the drainage system on the local hydrological cycles. However, in the case of the individual drainage systems the polluted water is treated near the source of pollution so people are bound to have a better awareness of the polluted water and its destination, and they have an opportunity to think about the local hydrological cycles.

But the problem of large-scale underground drainage doesn't stop here. Usually large-scale underground sewers carry water to a downstream treatment facility. As a result, in fine weather the upstream river water flow is greatly reduced and sometimes the rivers actually dry up. In contrast, individual drainage systems process water close at hand and discharge it into the rivers so that this does not happen. From the perspective of the local hydrological cycle, it is better to deal with rainwater drainage and treatment of sewer water close to where it originates. The processed water should be retained in the local hydrological cycle and clean water returned locally to the river. The rivers should not be allowed to run dry. In new urban development patterns, drainage should be clearly placed within the local hydrological cycle. As stated, it is important to strive for a change from large scale to small-scale drains based on individual drainage systems.

Second, the role and limitations of sewers should be clearly understood within the local hydrological cycles. It is true that sewers are needed to stop the pollution of river and lake water, but their use alone will not return a clear flow to the rivers. Although sewers in Tokyo are close to 100 percent of potential installation, the rivers within the city have not returned to their former clear flow. That is because the watershed's hydrological cycle has been destroyed and the original flow volume of urban rivers has been reduced. Of the water volume flowing through the Sumida and Tama Rivers in Tokyo, 66 percent and 43 percent, respectively, consists of treated sewer water. As the river's original flow is reduced there is virtually no way that the drainwater can be diluted; water quality is determined by the quality of the water discharged from the sewer treatment system. Because the quality of treated drainwater at best has a biological oxygen demand (BOD) of 10-20 mg/l, without any dilution effect it is difficult to attain water quality of below 5 mg/l, a desirable level for fish to survive. Although the river water is better than in the past, for the above reasons clean water is still unachievable. It is true that if a more sophisticated treatment system is introduced the quality of drainwater will improve, but as stated earlier, besides high energy and financial requirements, this approach produces a large amount of treatment sludge. Thus it cannot serve as an ideal solution for new urban development patterns. This indeed is the lesson to be learned from Tokyo. In order to restore a clean water to the rivers, there is a limit to what can be achieved simply by building bigger sewer systems. From the perspective of cycle capacity discussed below, the discharge of drainage water must be decreased, the hydrological cycle in the entire watershed must be recovered, and the abundant natural flow of water in the rivers must be restored.

Incidentally, in cases such with the cities of Kobe and Osaka, the upstream cities take in river water for their waterworks, the water is used in the city, treated and then discharged back into the river, and then goes into a downstream city's intake. This cycle of intake and discharge is sometimes repeated a number of times along a river. Obviously, if discharge from a city upstream is nearly untreated, water quality will be further degraded as it makes its way downstream. Accordingly, if the river's flow is not subject to at least some dilution, the quality of water intake for water works will get progressively worse. A classic example of this is the Kanamachi Water Purification Plant, the largest in Japan. It draws its water from the lower reaches of Tokyo's Edo River, where the problem of eutrophication results in a musty odor in tap water. If the musty odor were the only problem, intensive treatment with ozone and activated carbon would improve the water. However, that solution would require enormous inputs of energy and finances. It would also fail to address the fundamental issue of preventing pollution upstream.

A number of requirements exist in order to ensure a supply of safe and clean drinking water all along the flow of a river, and these must be implemented in an integrated way. First, as stated, the local hydrological cycles must be restored throughout the entire watershed in order to restore to the original abundant and sustainable flow and to increase the dilution effect. Second, urban development must be fit to the cycle capacity of the entire watershed. Third, planning should seek to establish small-scale systems based on individual drainage. Fourth, by restoring natural riverbanks, the capacity small and medium sized rivers to purify themselves must be strengthened. Fifth, through water conservation and multiple uses of discharge depending on water quality, the pollution burden on rivers must be reduced. This strategy is basically the same as on the Rhine River which crosses national boundaries in Europe, and supplies water and receives discharge numerous times as it flows.

3. THE NEW PARADIGM

3.1 Introducing Cycle Capacity

In the past, water was a limiting factor in the growth of cities as the size of a city was determined by the environmental space for water. Tokyo's original water supply was from small dams upstream along the Tama River. With that source alone the Tokyo one sees today could not have developed. However, in 1964 at the time of the Tokyo Olympics, Tokyo was assailed by a great drought, and this provided the incentive for Tokyo to shift its water source from the Tama to the Tone River system. It seems that this expanding of water supplies removed a limiting factor and accelerated Tokyo's concentrated development. But now this strategy has also reached its limits. Although a number of dams have been built on the upper reaches of the Tone River to supply Tokyo, every three or four years a water shortage occurs during the summer. This is partly because of a recent trend for less rainfall upstream in the watershed, but it also indicates that Tokyo's municipal strategy of combining massive water developments with neglect of the environmental capacity is gradually reaching its limits. Behind this situation are the unaddressed structural problems.

As previously explained, Tokyo planners must change their fundamental thinking about water policy. Further dependence on upstream sources must end. Tokyo must strive to become more water self-reliant based on the use of rainwater and groundwater close at hand. In order to do that the idea of environmental capacity must be brought into water resource policy. Also plans must be made to limit the urban concentration and thus prevent further large growth in the demand for water.

The factor of time must be brought into the idea of environmental capacity. As previously explained, rain is part of the local hydrological cycle and also sustains the local environment. Because time is an important factor for this local hydrological cycle, the local water's environmental capacity is really prescribed by both space and time. Although the idea of environmental capacity is not new, there was a tendency to consider it in a static sense. However, for sustainable development the dynamic aspect of 'environmental time' is necessary-the time that nature needs for a given cycle or to restore itself. Environmental capacity, which encompasses the concept of environmental time, is called 'cycle capacity' here. The concept of cycle capacity is not limited only to water problems, although this paper is limited to water; the concept is also an important basis for problems relating to natural resources and garbage and for a sustainable global environmental strategy.

Let us think of the use of groundwater from the perspective of cycle capacity. Rain seeps underground and over time becomes shallow groundwater; then over a very long period of time, it travels to deeper levels. Groundwater could be considered to represent the accumulation of both rainwater and time. Accordingly, sustainable use of groundwater requires the consideration of both the storage capacity for groundwater and environmental time. If these are neglected and groundwater is pumped up quickly, groundwater will disappear quickly. At present, near the Tama River water is being pumped up from deep subsurface layers to serve Tokyo. If large quantities of water are pumped up from the deep strata, a 'wringing out' phenomenon occurs and groundwater will seeps from shallow into the deeper layers. In fact present use of deep groundwater may already be drawing some water that came from the shallow layers. If the volume pumped is increased further, the shallow groundwater will dry up, and the restoration of spring water will become difficult. Tokyo already experienced the results of ignoring groundwater cycle capacity and excessive pumping, when ground subsidence of over 3 meters occurred in the eastern part of the city. In the future, the city must end its use of deep groundwater which has a very long cycle time, to shallow groundwater which has a relatively short cycle. Efforts must be made to plan for groundwater utilization consistent with the groundwater cycle of shallow layers.

At present, cities throughout the world are pumping up large quantities of deep groundwater because of its good quality and its low cost. But the rapid consumption of this deep groundwater while overlooking its cycle capacity risks irreversible contamination of this water with polluted water and saltwater, as many cities have already experienced. This is a very important problem as it applies to sustainable water strategies for new development patterns.

The next point to consider is purification of rivers from the point of view of cycle capacity. Japanese rivers of fifty years ago were clean flowing everywhere. Although some sewer water was discharged water pollution was not a problem because within the watershed as a whole there was a rich flow and a sufficient dilution effect, and the river's self-cleaning function was at work. The river's purification was fundamentally regulated by the dilution effect and by the river's power to clean itself. The former lay in the river's flow volume, and the latter in microorganisms living in the river. These microorganisms work with a remarkable combination of aerobic and anaerobic functions to purify sewer water. But for this to take place cycle renewal time, or environmental time, is necessary because of limits to the microorganisms' purifying capacity per unit of time. If the sewer flow is regulated within these limits, pollution of the river water will not worsen. However, in Tokyo, Osaka and many other large cities the population has increased rapidly within the watershed, discharging great quantities of sewer water, and resulting in rapid degradation of river water quality.

The main approach used today in sewage treatment is the activated sludge process, which utilizes aerobic microorganisms. From the perspective of environmental time, through efficient propagation of aerobic microorganisms as the main agents in the process, the time originally needed for purification can be shortened. This approach may provide one solution to today's pollution, which exceeds the rivers' present ability to purify themselves. Seen from the perspective of environmental capacity, the fundamental elements required to return a river to health include restoring the abundant hydrological cycle throughout the watershed, restoring the flow volume (environmental capacity), and with the original purifying capacity (environmental time). It is also necessary to control the volume and speed of discharge of sewer water. In order to accomplish all this efforts are needed to have human population levels appropriate to the watershed capacity, conserve water, use in stages depending on its quality (e.g., bath water could be used again to wash clothes), and promote recycling of sewer water.

3.2 Controlling Supply from the Demand Side

In addition to the concept of cycle capacity there is another important strategy for water in new development patterns: a complete change in approach to the supply and demand of water. Until now city planners have assumed that the future demand for water will inevitably continue to increase. However, from now the demand for water must be curbed. In other words, the supply of water must be controlled from the demand side.

The water supply and construction industries have traditionally overestimated future demand for water and have made an industry out of the excessive development of water supply projects and the area they covered. In order to recover costs through water rates there was no option but to encourage citizens and industry to consume a large quantity of water. Because over-consumption is encouraged when water is abundant, the appeals from authorities to conserve water during times of drought are unconvincing and generally have no effect. In the water industry one sometimes hears that if water conservation is thoroughly implemented and rainwater use is popularized, the sales of piped water will decrease, so their true sentiments are against water conservation devices and the use of rainwater. In addition, over-development of the water resource invites denser population and waste of water, which in turn induces new water developments; and so the vicious cycle continues. Tokyo is a classic example of this process.

As previously explained, cities have an inherent hydrological cycle capacity. In the future if we are to aim for cities that take into account the water cycle capacity, new water planning must be transformed from the current approach of supplying without controlling demand, to supplying based on controlled demand. Public bodies must fundamentally change to embrace self-reliant water resource policy. Also it is important to foster a consciousness of self-discipline with regard to water use by citizens and industry, based on the principle of providing for water needs using the supply available locally. If lifestyles based on mass consumption models which permit users to consume as much water as they like are not corrected, a sustainable new development pattern will be impossible. Lifestyle changes include the popularization of water conservation devices, but if citizens themselves do not switch to water conserving lifestyles, water demand will not decrease. It is interesting to note that citizens who have started to use rainwater report a transformation in attitudes from viewing rain as an unpleasant nuisance to seeing it as a blessing; they also report a greater interest in the distant dams that supplied them, as well as the issues of water use. Perhaps the change is thanks to having the water supplied from nearby, and it is exactly because they always use it that they develop a sense of discipline about water use.

According to Japan's Water Supply Law, the objectives of waterworks are to supply "clean" water at "low cost" in "abundance." However, now that the waterworks infrastructure has reached the point of almost one hundred percent saturation it is time for a change in policy. A fundamental reappraisal is needed of the objectives of the Water Supply Law from the perspective of the hydrological cycle capacity and suppression of water demand.

4. WHAT MUST BE DONE

Under a new paradigm introduced above, as cities strive for self-reliance of water supply and the revival of local hydrological cycles, practical issues such as the following will need to be addressed.

4.1 Institutionalization

First of all the storage, permeation into the ground and utilization of rainwater, as well as water conservation and recycling of sewer water should be codified into laws and regulations, and institutionalized so that they become a part of the fabric of society.

In 1982 Sumida Ward in east Tokyo requested the Japanese Sumo Association install rainwater facilities for water used in the National Sports Stadium. Based on this initiative's successful implementation, almost all public facilities in the ward were subsequently adapted to use rainwater. In addition, unique facilities for rainwater use known as rojison were set up by the efforts of local residents, making the use of rainwater a popular topic among both governmental personnel and citizens. After the two-year period starting in 1993, when Sumida Ward installed more rainwater use facilities than in the previous ten years, the ward carried out an analysis of the effectiveness of the use of rainwater in the city. It concluded that the use of rainwater had spread throughout the whole ward and that the rainwater facilities, known as "mini-dams," had been effective for irrigation, flood control and disaster prevention. It reported that based on certain assumptions, if rainwater utilization was expanded to thirty percent of facilities within the ward, the number of times that sewer water would be discharged into the river from the sewer pumps would be cut in half. In addition, in times of drought or disaster ward could supply eleven liters of water per person per day for about one month without relying on water from outside the ward.

Based on these results, in March 1995 the ward established "Rainwater Use Promotion Guidelines." The gist of the guidelines was first that in future rainwater utilization facilities should be installed at the same time as constructing new ward facilities. Second, developers of large-scale developments should be instructed and advised to use rainwater. Third, rainwater tank facilities for citizens should be subsidized. In October of that year, a subsidy programme began for rainwater tanks installed by citizens. By March of 1999, 114 rainwater tanks had been installed under the programme. At that time the total storage volume of facilities using rainwater in Sumida Ward had reached about 8,000 cubic meters, and rainwater "mini-dams" were continuing to grow in use throughout the ward.

In a place like Sumida Ward where the water table is high and the rainwater does not permeate far into the ground, the rainwater utilization system is based on accumulating rainwater and then using it. However, in areas where rainwater can permeate well into the ground, it is appropriate to consider water permeation into the ground, together with the use of water, from the perspective of sustaining the local hydrological cycle. In the future, water policies must take into account and institutionalize a combination of rainwater storage, ground permeation and utilization, making allowances for the local conditions.

In Tokyo, the average family consumes about 250 liters of water per day per person. In contrast to this, in Germany, about 140 liters are used, and in Berlin, an effort is being made to reduce this to 80 liters. If Tokyo residents managed to reduce consumption to the level currently consumed in Germany, they could achieve annual water savings of about 470 million cubic meters, or about 27 percent of the approximately 1.73 billion cubic meters of pipe water supplied in 1995. In the future, in order to achieve self-reliance in water local governments should switch to a demand suppression approach, which could include ordinances promoting water conservation and rainwater utilization, setting of clear objectives for the spread of both of these. In addition, policies could be introduced making it mandatory for not only public facilities, but also private facilities, to install rainwater utilization and water conservation equipment.

4.2 Incentives

To progress in making systematic changes in society, new incentives must be created. On this point government leadership is very important. It must not only show leadership through its own initiatives to promote self-reliance in water resources and the restoration of local hydrological cycles, but also support citizens' efforts and strive to integrate these approaches in society. Specific examples may include subsidizing facilities for using rainwater, devices for rainwater permeation into the ground, devices for water conservation or facilities for recycling sewer water, and reduction or exemption of charges for water utilities. Local governments that subsidize rainwater tank installation by citizens numbered 25 throughout Japan by August 1999, including not only Sumida Ward, but also Takamatsu City, Kawaguchi City, Tama City and Katsushika Ward. In Germany subsidies are also provided for facilities that help conserve groundwater, in addition to those that utilize rainwater. New York City has been successful by subsidizing water-conserving toilets. In Germany, some local governments have a system whereby they charge water utility fees when rainwater is discharged directly into sewers, allow an exemption of fees if rainwater is properly utilized. Such a system does not yet operate in Japan, but should be considered in the future.

4.3 Technology Development and Training of Technical Specialists

It is important to further develop technologies and train specialists in these fields. Technological advances are needed for inexpensive and efficient equipment to conserve water, facilities to use rainwater and devices to promote the rainwater permeation into the ground. Together with this is the need to train specialist with a thorough grasp of the technologies. At present the amount of water consumed in one flush of Japanese toilets is 12 to 13 liters. Compared with this Germany toilets flush only 6 liters the goal is 4 liters in the future. National and local governments should urge manufacturers to improve their technologies on products, which consume large amounts of water, such as toilets and washing machines. Governments should certify the best products and encourage their use. Similar initiatives could promote equipment that utilizes rainwater and promotes rainwater permeation into the ground. Fukuoka City established official measures and guidelines that encouraged water-conserving devices after an abnormal drought in 1978 and achieved success in popularizing them.

The German association "fbr", consisting of architects and other specialists in the use of rainwater, organizes seminars aimed at developers from various parts of Germany. The activities of this active group also include responding to design inquiries from the general public about facilities for rainwater use. In Japan there is not yet this kind of activity by people in the industry. There is now an urgent need for the development of groups of technicians well versed in design, construction and maintenance in the fields of rainwater management, the permeation into the ground, conservation of water, the multi-stage use of waste water, and water recycling.

4.4 Networking

Finally, in order for this kind of social system to continue in a sustainable way, networking among governments, citizens and industry within the country is important, in order to discuss together the key issues relating to water strategies. This is true not only of Japan but also throughout the world. In addition, in order to bring about sustainable new development patterns on a global scale, it is important to encourage regional exchanges and networking of these three groups internationally. Today Japan is carrying out advanced activities relating to rainwater utilization, ground permeation, and water conservation and it is hoped that in the future Japan will make a positive international contribution in these areas.

REFERENCES

Group Raindrop (Edit.). 1994. *Yatte miyo amamizu riyo* (Lets Try Using Rainwater). Hokuto Shuppan.

Murase Makoto. 1996. *Kankyo sigunaru - gemba de migaku kansei to kagaku* (Environmental Signals-Sensitivity and Science Honed on the Spot). Hokuto Shuppan.

Oshida Isao and Solar System Research Group. 1982. *Toshi no mizu junkan* (The City's Hydrological cycle). NHK Books.

Solar System Research Group. 1994. *Jyunkan toshi e no kokoromi* (Planning for a Cycle Town). NHK Books.

Tokyo Metropolitan Government. 1999. *Tokyo mizu junkan masta puran* (Tokyo Hydrological Cycle Master Plan).

Chapter 7

TOWARD SUSTAINABLE AGRICULTURE AND DEVELOPMENT

TAKESHI HARA
Waseda University, Tokyo

1. BACKGROUND

1.1 Definition of Agenda 21

Agenda 21, the Action Plan for Sustainable Development for humanity, adopted at the 1992 Earth Summit, defines in Chapter 14 the major objectives of "Sustainable Agriculture and Rural Development" (SARD) as being to increase food production in a sustainable way and enhance food security.

Agenda 21 suggests the necessary conditions for achieving these objectives, which involve educational initiatives, utilization of economic incentives and the development of appropriate and new technologies, thus ensuring stable supplies of food that meets nutritional needs, access to those supplies by vulnerable groups, and production for markets; employment and income generation to alleviate poverty; and natural resource management and environmental protection.

According to Agenda 21, the priority must be on maintaining and improving the capacity of the agricultural lands that have the greatest potential to support an expanding population. However, conserving and rehabilitating the natural resources on lands that has lower potential is also necessary in order to maintain sustainable population/land ratios.

1.2 The Concept of Sustainability in American Farming Practices and the Indian Green Revolution

In defining "the environment and sustainable development," there has been a fundamental difference between definitions of terms on what is "sustainability" in

K. Matsushita (ed.), Environment in the 21st Century and New Development Patterns, 163–197.

developed and developing countries since the first United Nations Conference on the Human Environment in 1972.

"There should no longer be a substantial difference between economists' and ecologists' goals. Our common goal should be the global protection of the atmosphere, oceans, soil and forests with a full reckoning of how human activity impacts the Earth's environment." (From a speech by Russell Train, the U.S. Ambassador to the United Nations, at the United Nations Conference on the Human Environment.)

Train's speech presented a new thesis, advocating the synthesis of economic activity and human interaction with the environment. It was a declaration of the movement toward "sustainability" made manifest by the people later called ecologists.

With respect to agriculture, this principle was put into action using the Low Input Sustainable Agriculture (LISA) proposed by the 1985 U.S. Farm Bill, which

Figure 7-1. Objectives and Methods of Sustainable Agriculture.

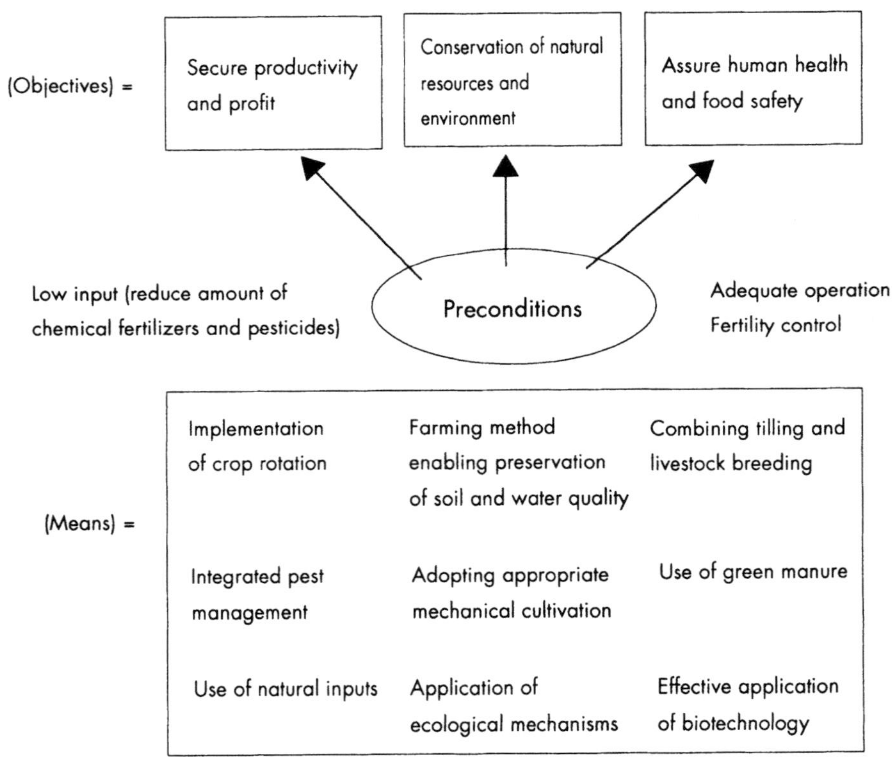

Source: Ryohei Kada, Environmental Protection and Sustainable Agriculture, *Ie no hikari kyokai*, 1990, p. 197.
Note: Adapted from U.S. Department of Agriculture, Economic Research Service materials.

was enacted during the agricultural recession. LISA refers to low input farming systems that optimize regeneration and renewal of resources and minimizes the use of commercial fertilizers and pesticides in order to maintain productivity and profitability while conserving natural resources and the environment as well as producing safer food. (See Figure 7-1.)

In developing countries, on the other hand, the main goal of their agricultural policies is represented by the "Green Revolution." It refers to an agricultural policy that makes it possible to meet the increasing need for major grain supplies that accompanies population growth by cultivating high-yield hybrid varieties of rice and wheat and increasing the yield per acreage.

For example, India launched its Green Revolution in 1967 and has been steadily increasing crop productivity. In Punjab state, where the land has been well irrigated since the days of British colonial, the rice yield per hectare increased from 1 tonne in 1967 to 3.5 tonnes in 1978.

The Green Revolution has been looked up to as a star of hope for farming industries in developing countries and has actually saved growing populations from starvation.

However, the introduction of exotic crop varieties to extensive farming, an agricultural system traditionally used in developing countries, transformed it into a high-input agriculture system that required irrigation and a large amount of chemicals and pesticides. For example, the amount of fertilizer put into the total arable land (140 million hectares, of which 15.4 million hectares were planted with high-yield varieties) was 2.17 tonnes in 1970-1971. These figures increased rapidly to 143 million hectares, 63.9 million hectares and 12.5 million hectares respectively. The use of agricultural chemicals lead to soil and water pollution and massive breeding of harmful insects that were tolerant to pesticides. The number of pest insects found in a field was three in 1965, which increased to eleven in 1990.

1.3 Shift from the Green Revolution to the "Ever Green Revolution"

The Green Revolution ran into a wall. Its methods conflicted with ecosystems and material cycles in developing countries. It strained societies in developing countries as the arising problems of unequal distribution and increasing expenses for irrigation systems and high-input farming become too much of a burden for small farmers.

Dr. Monkonbu S. Swaminathan, a former proponent of the Green Revolution in India, is now attempting to reverse the current situation by implementing the "Ever Green Revolution." Among the three principles of the Ever Green Revolution: Pro-poor, Pro-woman and Pro-nature, the principle of Pro-nature provides a basis for the on-going experiment of the bio-village, in which traditional Indian agricultural methods and indigenous crops are reexamined and integrated into modern agricultural methods. In a sense this experiment is an Indian version of the American

LISA and seeks a sustainable agricultural system. Dr. Swaminathan was invited by Premier Li Peng to join the International Committee for Sustainable Development organized by the Chinese government. The bio-village experiment is being conducted in China as well and its outcome is attracting attention.

The common goal of agricultural reform shared by the U.S. and India should be considered as the starting point for the development of a sustainable agricultural system that supports lives without destroying the environment.

The principles of LISA proposed in the 1985 U.S. Farm Bill were further expanded in the 1990 U.S. Farm Bill (The Food, Agriculture, Conservation, and Trades Act of 1990) which incorporated more up to date issues with consideration of the present situations in developing countries in order to deal with issues of environmental protection as well as business and the economy. It is highly instructive that the 1985 U.S. Farm Bill and the Indian bio-village project share the same principles aiming for sustainable agricultural systems.

1.4 Concepts of LISA in the 1990 U.S. Farm Bill

According to the 1990 U.S. Farm Bill the term sustainable agriculture means "an integrated system of plant and animal production practices having a site-specific application that will, over the long term: (1) satisfy human food and fiber needs; (2) enhance environmental quality and the natural resource base upon which the agricultural economy depends; (2) make the most efficient use of nonrenewable resources and on-farm resources and integrate, where appropriate, natural biological cycles and controls; (4) sustain the economic viability of farm operations; and (5) enhance the quality of life for farmers and society as a whole."

The 1990 U.S. Farm Bill addresses six items as preventive measures against environmental destruction. The six newly introduced measures are as follows.[1]

(1) Record Keeping for Pesticide Use

It became mandatory to keep records on the amount of restricted chemical pesticides used. Access to such harmful chemicals was limited to the federal government (the Department of Agriculture and the Environmental Protection Agency), state governments and medical personnel only.

(2) Water Quality Protection Program

The government provides subsidies in order to promote voluntary adoption of farming methods that reduce use of chemical fertilizers, which in turn protect water quality in the areas where agricultural production poses a threat to the quality of groundwater. The program set the maximum amount of a subsidy per farm as $3,500 and aimed to register 4 million hectares over five-year period.

(3) Marshland Conservation Program

This was a newly established voluntary program with federal incentives, aiming at protection of marshland. The federal government contracted with landowners for easement under thirty-year leases and the landowners were not allowed to change the usage of marshland without the federal government's consent during that period.

(4) Guidelines and Labeling for Organic Produce

Federal guidelines for organic produce were established and labeling on produce was promoted. The guidelines set some strict regulations: for example, produce was not considered organic if synthetically compounded fertilizers and pesticides were used in production; inspection results on pesticide residues were required for issuing a certificate.

(5) Sustainable Agriculture Research Authorized for $40 million per year

Under the 1990 U.S. Farm Bill, specific sustainable agriculture research was authorized for $40 million (about 5.6 billion yen) per year. This program was carried over from the 1985 U.S. Farm Bill. Considering that the amount authorized under the 1990 Farm Bill was far over the total amount of $8.3 million spent in the 1988-1989 fiscal year, the Bill dramatically demonstrated the federal government's will to promote environmental protection.

(6) Establishment of the Office of Environmental Quality

The Office of Environmental Quality was established in order to assess the impact of agricultural policies on the quality of environment. The Office was responsible for organizing and monitoring efforts to improve environmental quality.

(Note: Excerpted from Ryohei Kada, Environmental Protection and Sustainable Agriculture, Ie no hikari kyokai, 1990)

In some areas in Europe where environmental protection needs are high, the European Community (EC) in its practice of the environmentally sound agriculture guarantees a compensatory payment for loss to farmers who agree to engage in environmentally-safe farming for more than five years. Such agricultural practices include reducing use of chemical fertilizers and pesticides and growing grassland for maintaining the natural beauty of the landscape.

As it has been described, in their agricultural practices reducing the risk of groundwater contamination is emphasized in the U.S. and preservation of a beautiful landscape is highly regarded in Europe. Both aspects are equally essential in Japanese agricultural practices.

The Ministry of Agriculture, Forestry and Fisheries in Japan is finally going to submit a "Sustainable Agriculture Promotion Bill" to an ordinary session of the Diet in 1999. It aims at promoting the implementation of advanced production methods that will realize effective soil management and limit agricultural chemical use in order to establish a system of sustainable agriculture.

With all the current facts considered, it is expected that sustainable agriculture which strives for the coexistence of ecological and socioeconomic systems will be the fundamental basis for a new pattern of development in both developed and developing countries for the sake of environmental protection.

Table 7-1. Declining water resources as a result of deforestation.

Land Use	Forest	Terrace Farm	Slash-and-Burn	Pasture
Mm/hr	450	263	200	94

Note: Amount of water absorbed (mm/hr) per unit of area
Note: Comparative examination of the rainwater absorption of a pine forest located in the suburb of Baguio in central Luzon Island, Philippines.
(Surveyed by the National Forest Research Institute of Philippines)
Source: Science and Technology Agency (Japan), Basic Investigation on Devel

1.5 The Structure of "Non-Sustainable Agriculture" in Developing Countries

By the year 2030 the population of Asia, which now comprises 60 per cent of the current global population of 6 billion, is expected to exceed 8.5 billion. The population density in Asia is already high in most areas, with 200 to 400 people per square kilometer.

The rapid growth in population has led to overexploitation of croplands and excessive cultivation of slopes, resulting in topsoil erosion and loss of soil fertility, and indiscriminate deforestation for fuel wood. When a forest is cleared for farming or livestock grazing, usually by the slash-and-burn method, its ability to absorb rainwater declines greatly as shown in Table 7-1. When a forest loses its function in fostering water resources, it means a greater risk of water shortages and flooding in urban areas downstream. This was what caused the floods and poor harvests witnessed in North Korea during the late 1990s. Alarmed by a similar situation, the Thai government banned the felling of trees. The Chinese government, under siege of frequent floods, implemented afforestation projects aimed at reclaiming cultivated lands. Excessive use of natural resources has been the cause for topsoil erosion, desertification, and rapidly diminishing tropical forests and species. (See Table7-2 and Figure 7-2.)

Table 7-2. Expansion of Desertification.

Total ----------------------------------	more than 5 millions hectares/year
Irrigated cropland ----------	1 – 1.3 million hectares/year
Non-irrigated cropland ------	3.5 – 4 millions hectares/year
Desertification is occurring in large areas of pastures as well.	

Source: UNEP, "Status of Desertification and Implementation of the United Nations Plan of Action to Combat Desertification."

Figure 7-2. Ratio of human-induced soil degradation.
(Ratio of soil degradation area, due to over-pasturage and inappropriate agricultural processes, to total arable area)

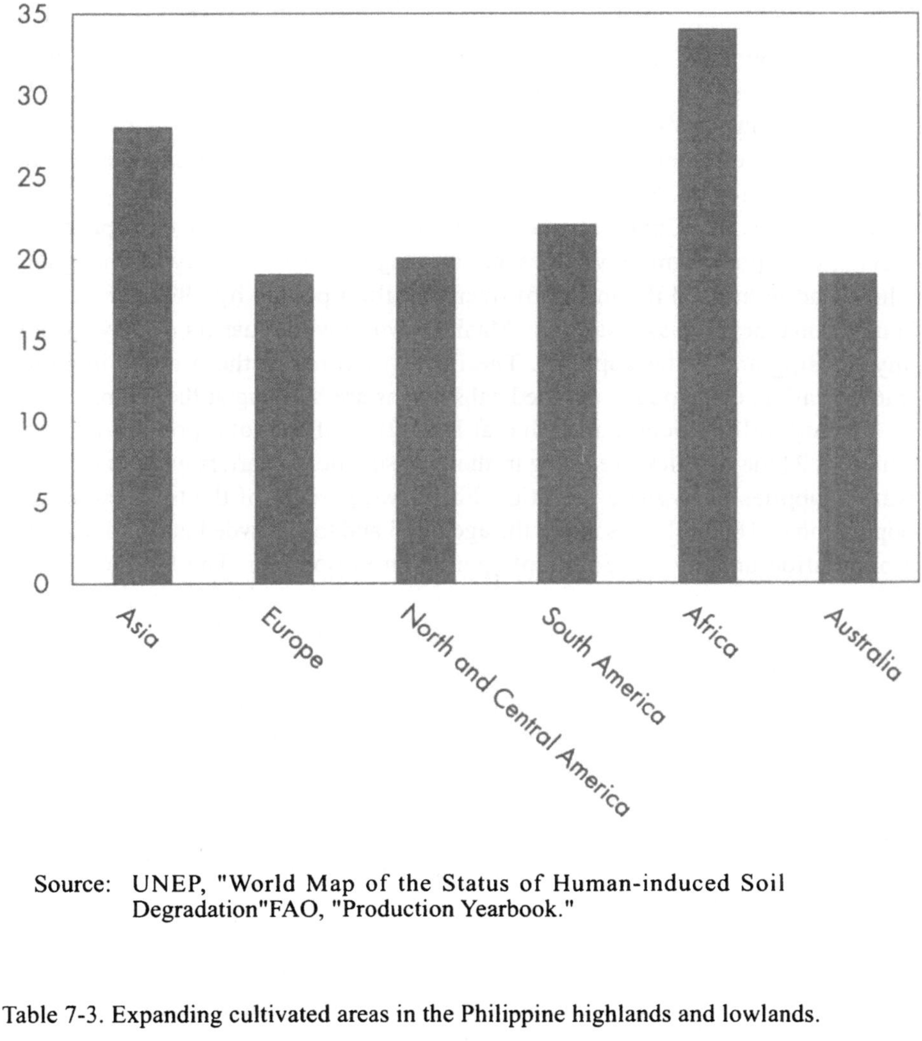

Source: UNEP, "World Map of the Status of Human-induced Soil Degradation"FAO, "Production Yearbook."

Table 7-3. Expanding cultivated areas in the Philippine highlands and lowlands.

Type of land	Cultivated Area (1,000 ha)				Annual Increase (%)		
	1960	1971	1980	1987	1960-71	1971-80	1980-87
Cultivated Area in Lowlands	5,581	6,424	7,760	8,845	1.3	2.1	1.9
Cultivated Area in Highlands	582	1,283	2,349	3,927	7.5	7.0	7.6
Total	6,163	7,707	10,109	12,772	2.1	3.1	3.4

Source: Maria Conception Cruz, Carrie A. Meyer, Robert Repetto et al., Population Growth, Poverty, and Environmental Stress: Frontier Migration in the Philippines and Costa Rica (World Resources Institute, Washington, D.C., 1992).

Treaties and conventions such as the UN Convention to Combat Desertification, the UN Convention on Biological Diversity, the Convention on International Trade in Endangered Species of Wild Fauna and Flora, and the Statement of Forest Principles are manifestations of the fact that the excessive use of natural resources has hit its limit. For example, the area of cropland cultivated in forests of the Philippine highlands was 580,000 hectares in 1960 but the area expanded to 3.9 million hectares by 1989. Soil erosion has become a serious problem. Topsoil of cropland is being gradually washed away by rain and wind at the rate of 122 - 210 tonnes/ha per year compared to the rate of erosion of only 2 tonnes/ha per year had it remained as forest. (See Table 7-3.)

. In countries like China, Thailand and Indonesia, a large number of people are leaving their poor farming villages and moving to cities. The city of Shanghai in China had witnessed the influx of over 2 million people by 1997. The current situation in cities such as Jakarta and Manila is worse with squatters who live without any housing and water supplies. The infrastructures of these cities including transportation, educational and medical systems are bursting at the seams.

Twenty million people in India, at least 28% of the total population of the country's 23 major cities, are living in slums or squatter's quarters without sufficient water, supplies or hygiene facilities. Forty-two percent of the total Indonesian population of 180 million is under the age of 15 and the crowded island of Java has a population density of 755 people per square kilometer. The United Nations Population Fund (UNFPA) analyzed the social structures of the "population influx phenomenon - urbanization" evident throughout all of Asia and identified the following possible causes:[2]

(a) Population increases in rural areas place an increased burden on food production.
(b) Increased agricultural investment in large-scale farming rather than on small-scale labor intensive farming resulted in a decreasing demand for labor, and thus caused people to migrate from farms to cities.
(c) Public policies emphasizing urban development fostered commercial and non-commercial activities as well as establishment of recreational facilities in urban areas rather than rural areas.
(d) Wages are higher in urban areas.
(e) Rural development emphasizing development of the infrastructure (such as building roads) improved mobility and facilitated migration to cities.
(f) Land distribution practices increasingly ignored traditional communal rights (such as livestock grazing and fishing). Those who didn't own land were left without prospects for making a living and moved to cities.

1.6 Perspective of Ecological Refugees

An examination of the root cause behind the migration of boat people to Japan may reveal an ecological factor, which motivated migration. Such an examination would reveal their status as ecological refugees. The increasing burden to feed a growing population resulted in deforestation, over-cultivation and overgrazing and eventually a limited agricultural yield. It was this vicious cycle of increasing population growth and the increasing burden on the land to sustain the growth through agricultural production that unleashed a flood of ecological refugees.

The author once visited the home town of a Thai family from northeast Thailand that had migrated to Japan and was living illegally in Tokyo, moving from one place to another. In their homeland, after complete deforestation of the area caused decreased precipitation, people lost their farms and became refugees. A song by a popular singer Et Karapao played on the radio...

I pray for rain when I wake up in the morning
When rain falls, I will remember
Rain does not come from the sky, but
It comes from the forests on Earth

The following sections will examine farmers' attempts to seek "sustainable agriculture" in India, Japan and the United States, taking into account the current status of agricultural production in both developed and developing countries.

2. STEPS TOWARD SUSTAINABLE AGRICULTURE

2.1 Bio-Villages in India

In Madras, the capital of Tamil Nadu in southern India, a hot wind comes off the Bay of Bengal blows through the streets of a town teeming with a crush of people and herds of cows, amid the black smoke of diesel trucks and clouds of dust. Here the author met Dr. Swaminathan, the father of Indian Green Revolution. The discussion centered on his visions of the "Ever Green Revolution" that seeks to attain increased food production to meet the needs of a rapidly growing population and create an agricultural system in balance with the natural ecological system.

Author: The global population is expected to reach 6,250 million by the year 2000. Whether or not the Earth can feed the mouths of an increasing population in developing countries is one of the primary concerns for Japan, the world's biggest food importer. You carried out the Green Revolution in India, a large nation both in its scale of population and agriculture. Tell me about your perspectives on global population and food supply using India as a model.

Dr. Swaminathan: When I first started the Green Revolution in India in 1967, the yield of wheat was 800 kilograms per hectare. We then worked on improving varieties and cultivation techniques and invested in irrigation facilities. As a result the yield increased by an average of three tons. But the yield of rice per hectare is not more than two tons, much less when compared to that of Japan, which enjoys a yield of five tons per hectare. It is possible to increase food production even more by shifting from extensive farming to an integrated management of cultivation and afforestation using abandoned croplands and adopting crop varieties and cultivation techniques that have been proven productive.

Author: It is still necessary, however, to increase food production by 30% in order to provide the Indian population which is increasing by at least 7 million a year with the adequate calories needed to meet international standards.

Dr. Swaminathan: Before I give you a statistical answer to your question on how many people we can feed, I would like you to take a closer look at the situation of India. It is possible at the very least to satisfy the food needs of a large population. But how can people increase the quality of their lives and bring joy to their existence? What about housing, education, medical care and employment? Even if food production is increased, we are not sure if poor Indian people can afford to buy food, as is the similar case in other developing countries.

Author: The Green Revolution saved people in India from starvation. However, growing high-yield varieties required the use of large amounts of chemical fertilizers and pesticides and caused the breeding of new types of pests that are resistant to pollution and pesticides. The variety of crops traditionally grown was limited to high-yield varieties only and it seems that the basis for a secure and sustainable food source has been destroyed. Can we consider the Green Revolution that became a global agricultural model to be a kind of "sustainable agriculture"?

Dr. Swaminathan: The Green Revolution certainly had some drawbacks behind its success. But there was an urgent need for starting the Green Revolution in this poor farming society where food, employment and income were people's primary concerns. It was the society that made Mahatma Gandhi lament: "for those who are hungry a loaf of bread itself is their god and those who do not have clothes a strip of cloth itself is their god."

Author: Tell me about the "Ever Green Revolution" that your research institute is leading in an effort to make up for the failures of the Green Revolution.

Dr. Swaminathan: It is an agricultural practice that seeks to coexist with nature and to bring profits to the poor and women, in other words, it seeks to accomplish social equality. In an experimental Bio-village with 450 households located near Pondicherry, which used to be a French territory, an "Eco-technology Movement" has been started, seeking maximum production with minimum resource consumption. This is an attempt to grow various kinds of traditional crops, not just a single variety, combining traditional organic farming methods with modern biotechnology and using solar and wind power and biomass as an energy source. We are making progress as the creation of employment opportunities are realized in accordance with the principles of ecology and equality. We have cooperated with China and Guyana, countries that are both facing similar population and food problems, and bio-villages are being created there. UNESCO and the Cousteau

Foundation in France have started a network for international support. I believe the "Ever Green Revolution," learning from the failures of the Green Revolution, will open the way to sustainable agriculture and be the starting point in sustaining hopes for human society in the twenty-first century.

The Ever Green Revolution proposes the concept of eco-technology. It's principles are expressed by the "three Es," Ecology, Economy and Equity, and it aims for ecological conservation and the realization of sustainable agriculture. These are working principles for the bio-village in Pondicherry where the Ever Green Revolution is practiced.

(a) Without ecological sustainability there will be no lasting profitability.
(b) Without economic potential there will be no technological development.
(c) Without economic equality between genders the poverty gap will widen and there will be no societal unity.

In order to fulfill the purpose of these three principles, the Integrated Intensive Farming System (IIFS) has been implemented in the bio-villages. The IIFS proposes three objectives in an effort to offset the negative impact posed by the preceding Green Revolution.

(a) To achieve higher yield without harming the environment
(b) To disperse demand on specific agricultural products by diversifying diets
(c) To enhance farmers' profitability by increasing the value of agricultural products by using post-harvest processing technologies.

The three mottoes, Pro-nature, Pro-poor and Pro-women emerged from this background. What is noteworthy is that Dr. Swaminathan, based on his own experience, highly regards the principle of Pro-women. Although in India most women are the poorest of the poor and in most cases illiterate, he sees a great potential in them. In bio-villages women living in the poorest level of the social class mastered mushroom production techniques and passed them on to many other women. Dr. Swaminathan has high regard for this accomplishment and sees it as an economic success in which, as Mahatma Gandhi once addressed in his teachings,

Table 7-4. Excess Rate of Nitrogen Acid in Public Water and Groundwater. (10mg/liter)

	Number Inspected	Number Exceeded	Rate of Excess(%)
Rivers	2878	4	0.1
Lakes and Marshes	660	0	0
Seas	1145	0	0
Groundwater	3664	168	4.6

Source: Environment Agency of Japan (1995-1996).
Note: This is not a nationwide study.

economic growth is promoted through production by the individual, creating more employment opportunities, not by mass-production, which promotes mass unemployment. In another case, some women succeeded in producing hybrid seeds and sold them to companies, resulting in an added value to their time and labor. Dr. Swaminathan gives high marks for these accomplishments and sees these activities as paving the way to a new style of sustainable agriculture in farming villages throughout developing Asian countries.

2.2 Initiatives in Japan

2.2.1 The Environmental Load and Problems of Material Flow in Japanese Agriculture

"Many cows these days are becoming weak, they cannot bear a labor of giving birth," said one dairy farmer, grieving changes in dairy farming in Nemuro,

Table 7-5. History of Environmentally-sound Agriculture in Japan.

The terms "environmentally-sound agriculture" and "sustainable agriculture" are used synonymously here. The history in Japan is shown below.

1948	Mokichi Okada, now a leader of Sekai Kyusei Kyo (World Saving Religion), presented a paper on farming without pesticides.
1953	Yamagishi-kai was organized.
1959	*Agriculture Bible* by Howard was published.
1971	Japan Organic Farming Research Association was organized.
1974	*Fukugo osen* (Complex pollution) by Sawako Ariyoshi was published serially.
1985	Natural Farming International Research and Development Center was established.
1987	The Liberal Democratic Party organized the Federation of Organic Farming Research. Specially cultivated rice system was established.
1988	Ecosystem Agriculture Liaison Council was organized. The Socialist Party organized an organic farming research group. The Central Union of Agricultural Cooperatives hosted the first National Organic Farming Agricultural Cooperative Exchange Meeting.
1989	The Ministry of Agriculture, Forestry and Fisheries established the Organic Farming Policy Council within it's Agricultural Production Bureau.
1992	A new policy emphasizing environmentally sound agriculture was announced. The Organic Farming Policy Council established in 1992 was reformed and renamed the Environmentally Sound Agricultural Policy Council, which aims to promote environmentally sound agricultural practices such as reducing use of agricultural chemicals and fertilizers.

(Compiled by Nobutaka Suzuki, Director, Crop Production Division, Agricultural Production Bureau, Ministry of Agriculture, Forestry and Fisheries)

Hokkaido, Japan's prime dairy production region. In this area 200,000 cows are raised, over double the human population of 90,000. The excrement of one cow equals that of 50 humans, making the cow manure produced in the area equivalent to what Tokyo's 10 million people would produce. Although every dairy farm is equipped with a large manure processing facility that looks like an oil tank, costing each farm 40 million yen, only manure from cowsheds is collected.

The processed manure is applied as fertilizer to pastures. The main component of such manure is nitrogen, which becomes oxidized and turns into nitric acid, which eventually permeates into the soil and ends up in river water. As a result the river water and groundwater cows drink is contaminated by nitrate nitrogen (NO_3-N). Once taken into an animal's or human's body, nitrate nitrogen (NO_3-N) becomes deoxidized as nitrite nitrogen (NO_2-N). When combined with hemoglobin, it can cause a hypoxic reaction and lead to blue baby syndrome. The amount of nitric acid contained in water must be less than 10ppm under the Water Act. Due to hypoxia cows get short of breath during labor.

In Japan where the self-sufficiency rate of grain is 30%, more than 90% of feed grains are imported, amounting to 40 million tonnes a year, four times as much as the amount of domestic rice consumption of 10 million tonnes a year. Nitrogen is the main element in carbohydrates, the principal constituent of grains. The "imported nitrogen" accumulating in Japanese land is the highest in the world and its outrageously high amount totals 130 kilograms per hectare annually. Furthermore, nitrogen input in Japanese agriculture is 13.7 kilograms per hectare, far more than the world average of 5.4 kilograms. Because of this, nitrate nitrogen ranging from 0.5ppm to 3ppm is detected in groundwater in many places throughout Japan. (See Table 7-4.) Additionally, a large amount of sulfur and nitrogen oxide is emitted into the air due to the use of petroleum, another imported resource, causing sulfur and nitrogen to fall on land in the form of acid rain and dust. Its amount is 53,000 tonnes in Tokyo and 42,000 tonnes in Osaka (1985). Japan is now so nourished that it suffers from being "nitrogen overweight." If you fly along the shoreline from Haneda to Fukuoka in summer, you will find red or blue spots in the sea. This is the sight of red and blue tide caused by the abnormal breeding of plankton due to excessive amounts of nitrogen and phosphorus.

The Japanese government's White Paper on the Environment in 1994 pointed out in its introduction the excessive amount of nitrogen existing in Japan and warned that the material balance, a calculated flow of materials based on economic activity, has tipped against the environment.

How can the current situation in dairy farms in Hokkaido be improved? Some argue that instead of raising so many cows domestically, importing dairy products may alleviate the problem. However, as the new policy proposed by the Ministry of Agriculture, Forestry and Fisheries suggests (i.e., new steps toward food, agriculture and rural communities), the foundation for the Japanese dairy industry must be preserved in order to maintain self-sufficiency.

In order to reduce excessive nitrogen caused by animal waste from dairy cows, nitrogen can be circulated in the form of manure to other farming areas growing rice, fruit and vegetables, and be used to maintain soil fertility by using farm resources through integrated operation of crops and livestock. Nitrogen fertilizer inputs could be reduced and again nitrogen balance could be obtained.

It is necessary to establish a new agricultural system for restoring soil fertility and a distribution system for certifying organic farm products. It is expected that "guidelines for organic crop labeling" developed by Ministry of Agriculture and Fisheries will be put into use soon. The problem of excessive nitrogen must be solved much more than the Japan's trade surplus. For this purpose, Japan's agricultural style itself must shift from that of environmental destruction to environmental protection. Therein lies the way for Japanese agricultural industry to achieve excellence, public support and a responsible market.

The crop rotation system is one of the easiest and most practical farming methods for increasing profitability and reducing excessive nitrogen inputs. A model crop rotation pattern would be as follows: the year after planting *daikon* radish, a high user of nitrogen fertilizer, plant potatoes, not a high user of nitrogen fertilizer. In the following year plant groundnuts, low users of nitrogen fertilizer. In this way the nitrogen content in the soil balances out. Such a crop rotation scheme, aiming at environmental safety and preservation of soil fertility, is not welcomed in the modern distribution system that encourages crop rotation of certain kinds of identical crops in the entire farming community. This market mechanism is an obstacle and a disturbance to the environmental cycle of materials. Both producers and consumers must work toward reform of the current distribution system, which includes the wholesale market where farm products are treated as if they were standard industrial products.

2.2.2 Cooperation of Producers and Consumers: an Attempt in Rice Farming

Now that the Food Control System is no longer in effect, there is some resistance against opening the market to farm products. One example of environmentally sound agriculture that demonstrates its potential is "the joint development of new rice varieties." This was a joint venture undertaken by the Seikatsu (or life) Club Union (Chaired by Nobuhiko Orito), a co-operative network of 220,000 members in metropolitan areas, and the Japan Agricultural Cooperative (JA) in Yusa, Akumi, Yamagata Prefecture, a rice farming district along Gekko River at the foot of Mt. Chokai. (Refer to the interview with Nobuhiko Orito, Chairman of the Seikatsu Club, appearing later in this paper.)

This joint program employs rice-farming practices "desirable" for both producers and consumers in terms of safety, taste and productivity. The Yusa JA and the Seikatsu Club had worked on cultivating varieties of rice crops through trial and error for six years and finally found a cultivating method for producing good quality rice that both parties agreed upon. In 1993 they sold 1200 tonnes (20,000 bales) of the jointly harvested rice.

Based on the idea that the superior quality of nature is a fundamental requirement for harvesting high quality rice, the Seikatsu Club invested 12 million yen for restoring water quality in the area, in line with the Water Quality Restoration Act, a newly established ordinance by Yusa town. As a result, the water purity was restored, bringing back whitebait fish to the rivers near the Sea of Japan. The cost of protecting the environment is included in the price of the rice, which is determined through agreement by both rice producers and consumers.

After going through many tunnels under the Oou mountains, the Tohoku Yamagata bullet train arrives at Takahata Station. The town of Takahata, located in Higashi Okitama, Yamagata Prefecture (population of 27,500) is surrounded by mountains of Zao, Iide and Azuma in east, west and south and Mogami River runs through to the north. The rolling landscape varies in heights from 210 to 1,020 meters above sea level. It is a beautiful place the local people call *Mahoroba no Sato*. (*Mahoroba* is an old Japanese word for a beautiful landscape covered with mountains or hills. Sato means a country or a home town.)

Kanji Hoshi, chairman of Takahata's board of education, is a leader of rice farmers and a well-known farmer poet. In 1954 Hoshi took over his family farm and became a member of the Tohoku Rice Association (or *Kome no Kai*), that was promoting farm collectivization and mass-producing technologies. In 1961, the year after the Twofold Income Increase Plan was implemented, the Agricultural Basic Law was enforced, aimed at increasing profitability in farming industries, which was an agricultural version of the governmental policy to increase the growth of the Japanese economy. As a member of the Tohoku Rice Association, Hoshi was in the vanguard of the movement toward "profitable farming." In Takahata like any other farm area, rice farmers replaced their cows and horses with tractors and rice-planting machines, started using chemical fertilizers and pesticides and burned rice straws, once a main resource for a soil fertilizer but no longer used. Many households became more affluent, with home electric appliances and expensive imported liquors.

In 1972, Hoshi began to notice something was wrong: the apple orchard he had spent the past ten years nurturing was completely damaged by disease. He realized that the soil had lost its fertility over the years. In the following year he organized a Takahata Study Team on Organic Farming with 38 other young local farmers (average age 27). Their slogan was health, safety and environment. They used a cultivating method that did not use any agricultural chemicals, and the overall yield went down by 20-30%. Although their activities were not accepted by other farmers, they continued their effort to learn new organic farming methods and eventually achieved an increased yield, while contributing to environmental conservation. Organically produced rice crops were found not to be very susceptible to frost damage and started to gain good reputation among farmers in this cold region.

Their activities were also introduced in Sawako Ariyoshi's novel, *Fukugo Osen* (Complex Pollution) in the Asahi newspaper and met with a positive public response and gradually gained support among both farmers and consumers. Their fundamental principle of this approach is to increase and then maintain soil productivity through the elimination of chemical use and the reliance on organic substances such as manure.

Another organic farming group in town, the Kamiwada Organic Rice Producers' Association, was formed later and their objectives were to (1) advocate rice farming as an industry, (2) promote an understanding of the real nature of food and preserve natural qualities of the environment, (3) produce safer and better tasting high-quality rice, and (4) seek exchanges between urban and rural areas through rice production in an effort to ensure the continued existence of the town.

The association members are prohibited under the association regulations to use any type of agricultural chemicals and instead use whole-rice vinegar, but they are permitted to use a rice cultivation method with reduced agricultural chemical inputs which allows single application of herbicides. Most commonly, however, members use organic fertilizers that they have developed themselves. Such fertilizers are made from ground fish and bones, rice bran, residue from cooking oil, animal meat and fish soup stock, with added minerals.

Takahata farmers currently produce organic rice and vegetables as well as apples, grapes and pears without using any agricultural chemicals, while at the same time raising hogs and chickens without using synthetically produced livestock feed. "Takahata brand" farm products and processed products are now one of the most popular and highly priced items in the Tokyo market. Since the Food Control System was abolished in 1996 and the Japanese agricultural policy has shifted from governmental price control to market centered pricing, the agricultural practices initiated by the "Takahata farming" method and the joint program by JA and Seikatsu Club co-op are now gaining popularity and their economic success should serve as a model for the Japanese agricultural industry in the twenty-first century, aiming at free competition based on individual originality and ingenuity.

Today, 26 years after Hoshi first started his drive toward organic farming, the Takahata Study Team on Organic Farming (with 46 members) and Kamiwada Organic Rice Producers' Association (with 110 members) produce 6,000 bales of rice every year and sells them to consumer groups nationwide at 33,000 yen per bale (60 kilograms). Compared with the highest price of independently distributed rice, about 26,000-28,000 yen, the high price of Takahata rice indicates how highly it is esteemed by consumers.

Hoshi considers that the underlying principle for organic farming is love for life. Caring about every life on Earth and living with them is what organic farming is all about. In autumn Hoshi's paddy fields are filled with dragonflies dancing in the sky and diving beetles and whirligig beetles waddling in the water. In July 1988 the National Research Institute for Pollution conducted a test on the survival rate of

Figure 7-3. Implementation of Environmentally Sound Agriculture by Municipalities. (1995, multiple responses)

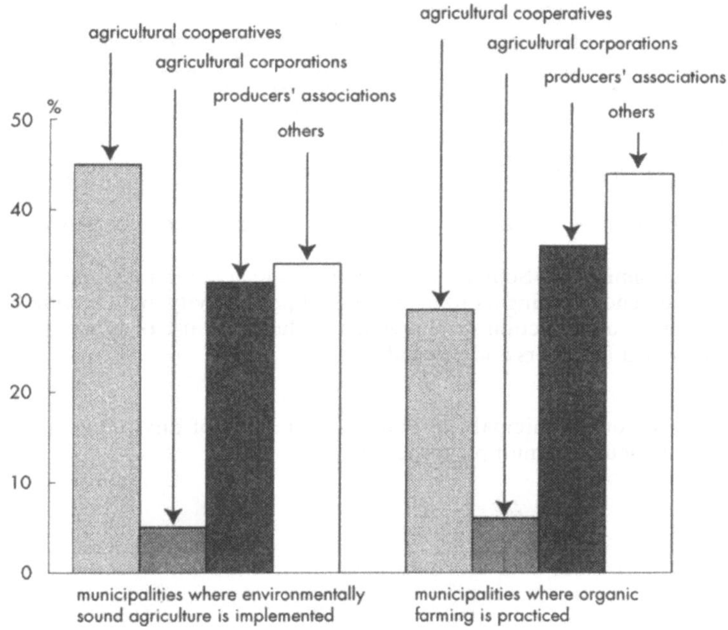

Figure 7-4. Distribution Systems in Environmentally Sound Agriculture. (1995, multiple responses)

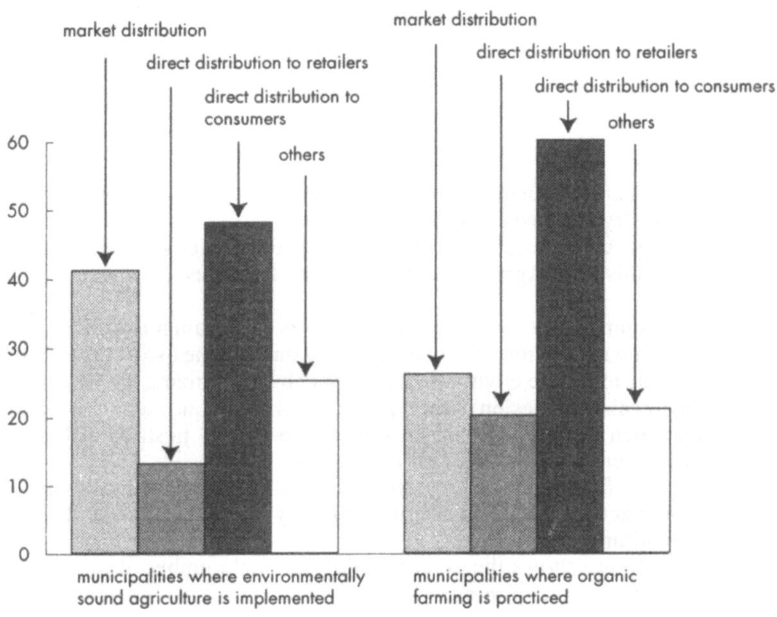

Table 7-6. Status of Implementation of Environmentally Sound Agriculture.

	Number of municipalities (ratio against total number of municipalities in %)	Breakdown by category (multiple responses)		
		JA	Corporations	Associations
Environmentally Sound Agriculture	2164 (19.0)	977	119	680
Organic Farming	897 (7.9)	257	58	304

Source: "1995 Census of Agriculture," Ministry of Agriculture, Forestry and Fisheries.

Note: Environmentally Sound Agriculture refers to agricultural practices using reduced chemical inputs to maintain soil productivity and Organic Farming refers to a production system which excludes the use of synthetically compounded fertilizers and pesticides.

Figure 7-5. Reasons Municipalities Support Promotion of Environmentally Sound Agriculture (multiple responses)

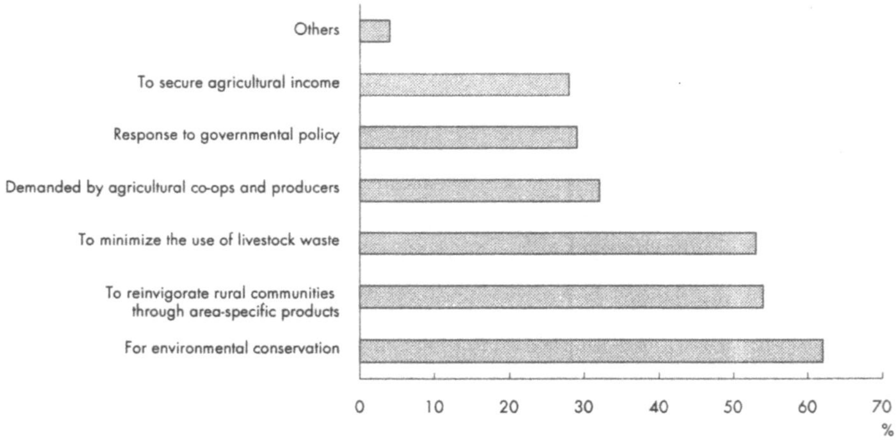

Source: "Report on environmental initiatives by farm communities, " Ministry of Agriculture, Forestry and Fisheries (September 1994)
"Census of Agriculture" (A comprehensive study on environment in agricultural areas), Ministry of Agriculture, Forestry and Fisheries

Note: (1) Environmentally sound agriculture refers to sustainable agricultural systems based on restoration of soil productivity through the use of organic substances. It seeks to reduce environmental impact by minimizing the use of agricultural chemicals. It relies on natural pest control techniques whereby natural predators such as wild fowl are encouraged to control pests as well as the use of pheromones to ward off pest infestation.

(2) Organic farming in this research refers to an environmentally-sound agricultural practices that eliminate the use of any types of synthetically compounded agricultural chemicals.

(3) Each figure shows the percentage of the total number of municipalities that practice organic farming.

Table 7-7. Reasons for Implementing Organic Farming. (Multiple responses)

	Total	
	Number of responses	Percent of total responses
Valid responses	1231	100.0
Health concerns due to the use of chemicals	869	70.6
Interest in cultivation techniques that prevent potential hazards of continuous cropping and loss of soil productivity	493	40.0
Access to organic manure	433	35.2
Opportunity for optimizing original production techniques	372	30.2
Environmental concerns in agricultural industry	592	48.1
Recommendation by governmental functions	115	9.3
Economic profitability	494	40.1
Regeneration of rural communities	327	26.6
Recommendation by other organic farmers	168	13.6
Demand from consumers and consumer groups	305	24.8
Recommendation by agricultural cooperatives	115	9.3
Others	119	9.7
No response	15	1.2
Total number of reasons selected	4402	—

Source: 1994 Tokyo municipal government report on organic farm product distribution and environmental governance.

a freshwater shrimp *Paratya compressa improvia* in Takahata. After four days from their release in water, all of them lived in Hoshi's paddy field but none survived in paddy field where pesticides were sprayed.

Hoshi was elected as a chairman of the board of education in Takahata and started incorporating rice cultivation in elementary school and junior high school education, based on his belief that both agriculture and education should be intertwined at the root to ensure the proper cultivation of young lives. The lifestyle of Hoshi and his community provided the basis for the big hit animation movie "*Omohide Poroporo*" (Memories drop by drop) (1991) and the NHK television documentary "*Mada minu tsumatachi e*" (To the wives we haven't met) (1992). Since then many young people touched by the movie and the TV show have visited the town, and 23 of them have settled there. Furthermore, over 3,000 college students come to the town and stay for training every year. Besides achieving economic growth, local farmers gained pride and confidence in their work thorough interaction with city dwellers. "Takahata Kyosei Juku" was opened in a house with a thatched roof in the village of Wada along with the "Mahoroba Farming School," with the

Table 7-8. Reasons for Purchasing Organic Farm Products. (Multiple responses)

	Total	
	Number of responses	Percent of total responses
Total number of responses	1080	100.0
Organic farm products are good for health	851	78.8
Good for the environment	408	37.8
Better tasting	241	22.3
Fresh	182	16.9
Safe	760	70.4
Nutritious	235	21.8
Available in stores	14	1.3
Better than other farm products for some reasons	288	26.7
Recommended by others	8	0.7
Talked about recently	78	7.2
Other reasons	26	2.4
No response	26	2.4
Total number of reasons selected	3091	—

Source: 1993 Tokyo municipal government's report on organic farm product distribution and environmental governance.

hope to promote the coexistence of nature and man as well as cities and farming communities.[3] They are similar to the "Rasuchijin Kyokai" established by Kenji Miyazawa in 1926 which expected to improve the culture of farming communities. Over one hundred people from Tokyo and Osaka have participated in programs at these schools.

Hoshi, a taciturn assistant juku-master at one school, sings.

My father, an expert rice farmer, smiles
Standing on a path between rice fields over there
In autumn you will see golden waves here
Don't let this scenery be taken away
Bring back the harvest lights
On this land of fertility
Harvest our beloved rice
("Ine no michi kou")

Under the old food control system, besides the two major distribution systems of independent and governmental rice, there had been a third distribution system established in 1987, in which individual producers and consumers could sell and buy rice with the approval of the food control office. The rice distributed in such a way is called "specially cultivated rice." In many cases such rice is grown organically or with reduced chemical inputs and is marketed to consumers in cities through individual efforts by farmers or through consumers' effort to find rice that meets their standards. The rice is distributed directly from producer to consumer, though it requires an approval from the food control office. The distribution of specially cultivated rice is significant in that it reflects consumers' demand for a drastic reform in the Japanese rice farming system that had always been controlled by the government, in terms of the production, distribution and consumption of rice.

The organic rice cultivated in Takahata and Yusa has its roots in the specially cultivated rice marketed by individual farmers. The environmentally sound agriculture that promotes organic farming practices is becoming widely recognized in many places in Japan. As indicated in Figures 7-3, 7-4 and Table 7-6, agricultural practices in which emphasis on quantity over quality allows farmers to use high chemical inputs are shifting toward the environmentally sound agriculture, seeking sustainable development for the future of Japan. At the same time, the organic farming system is becoming economically successful due to its potential profitability and increasing demand from consumers. From the perspectives on the nature of market economy, farmers alone should not be responsible for the effort toward sustainable agriculture, because sustainable agriculture stands on the economical success of products produced through sustainable agriculture.

As an example of a successful effort toward sustainable agriculture, a joint project by the Seikatsu Club co-op and Yusa JA in Yamagata will be described in the following interview with Nobuhiko Orito, Chairman of Seikatsu Club Union (headquartered in Shinjuku, Tokyo). The Seikatsu Club is a group of consumers, that consists of 235,000 members from 14 prefectures, and is the largest rice consumer group in Japan, purchasing annually 12,000 tonnes of jointly developed rice directly from farmers.

Author: According to the public survey on food, agriculture and role of rural communities conducted by the Prime Minister's Office in September 1996, 70% of respondents living mostly in metropolitan areas have concerns about the food situation in Japan. Reasons for such concerns included poor harvests due to abnormal weather and natural disasters as well as changing environmental conditions and population growth. Some people say that we should increase production of domestic products even if they cost more than imported products.

Orito: Considering the current rice situation is a result of the new food law, we must raise the question of why such issues of public concern are not addressed in governmental policies.

Author: The Seikatsu Club co-op is known for its joint project of rice cultivation with Shonai Midori JA in Yusa, Yamagata prefecture. Why should this project be considered as a model for jointly produced rice by producers and consumers?

Orito: In 1996 we purchased 60,000 bales at 21,120 yen per bale. It took us six years to develop this rice before we could satisfy both producers and consumers in

Figure 7-6. Roles of Rural Communities.

<Major Roles>

Provide farm, forest and marine products	Farm products	
	Forest products	Timber, special forest products
	Marine products	
	Energy	Hydroelectric power generation, geothermal generation
Provide living space and employment	Housing	Quiet and spacious dwelling
	Land for facilities	Factories, offices, recreational facilities
Land conservation	Prevention of degradation	Soil degradation, topsoil erosion, wind damage
	Prevention of natural disasters	
	Conservation of land and seas	
Preservation of water resources		Water storage, water control, water purification
Conservation of natural environment	Natural landscape	
	Climate control	Adjust atmospheric and terrestrial temperatures
	Air purification	Absorb CO2, provide O2, purification
	Wildlife protection	
Provide natural and cultural resources	Education	Learning based on research and experience, through staying in a rural community; cultivating aesthetic sentiments; sightseeing
	Recreation	Sports, outdoor recreation, marine recreation, wellness, community interchange
	Rural landscape	
	Culture	Protect cultural heritage

Source: Ministry of Agriculture, Forestry and Fisheries

terms of safety, taste, harvesting methods and productivity and then finally start selling it in 1993. When determining the purchase price, both consumers and producers get together at a meeting of the Jointly Developed Rice Department of Shonai Midori JA and decide on a price based on the production cost and other expenses. We also set aside a separate fund for the jointly developed rice in case of natural disasters such as cold weather damage. The rice is harvested without using herbicides and with limited use of pesticides.

Table 7-9. Estimated Value of Paddy Fields According to Substitutive Cost Method. (billion Yen)

FUNCTION	BENEFITS	ESTIMATED VALUE		
		Paddy Field	Other Fields	Total
Preventing Floods	Mitigating the damages caused by floods	1,952.7	388.1	2,340.8
Developing Water Resources	Supporting steady water flow and inexpensive ground water supply	739.8	23.6	763.4
Preventing Soil Erosion and Landslides	Mitigating the damages caused by soil erosion and/or landslides	47.2	5.5	8.252.7
Soil Purification	Reducing the cost of waste treatment	4.5	3.7	8.2
Preserving Rural Landscape and Recreational Amenities	Social interchange	1,711.6	1,458.1	3,169.7
Air Purification	Absorbing contaminants and purifying air	171.7	146.5	318.2

Source: The values in this table are based upon "Research on the Environmental Externalities of Paddy Fields" (Mitsubishi Research Institute, March 1991) and estimated with consideration to 1993 data provided by Ministry of Agriculture, Forestry and Fisheries. (1994)

Figure 7-7. Agriculture and Environment.

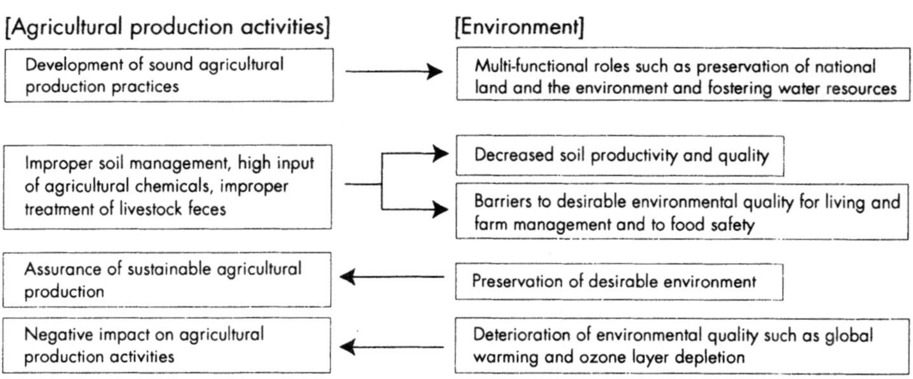

Source: Ryohei Kada, Environmental Protection and Sustainable Agriculture, Ie no hikari kyokai, 1990

Author: What about the social cost of rice farming?

Orito: The town of Yusa is located along Gekko River. Seven years ago the prefectural government gave permission for the construction of an aluminum recycling facility in the upper reaches of rice farming areas. For us rice and water mean the same thing. Pure water is indispensable for growing safe and tasty rice. So we raised money from members, collected 16 million yen plus supplemental funds paid by the town and used this money to move the facility to an industrial complex in Sakata. The town then issued an ordinance for restoring water quality. The entire community and even consumers in cities participated in this campaign. The river water became so clean that whitebait came back to the river from the Sea of Japan. A pure natural environment is essential to harvesting pure-quality rice. The cost for protecting the environment is the social cost of rice farming. It is our belief that rice consumers are responsible for the cost.

Author: Now that the Seikatsu Club is so influential in the market, purchasing 12,000 tonnes of rice annually, do you consider the possibility of launching another rice cultivating project that will affect pricing of rice crops?

Orito: It is our responsibility as a co-op to manage rice supplies. Our club members make a 14-month contract to buy rice. Individual members are supposed to store a two-month supply of rice no matter what circumstances they may be in. They must then mix 10% of the stored surplus rice with the new grain they receive. This system is for fulfilling consumers' responsibility to ensure a steady supply of rice, from its production to distribution.

Author: The Food Agency is having difficulty in selling 3 million tonnes of surplus stored rice. According to the new Food Act, 1.5 million tonnes of stored surplus rice cannot be sold for livestock feed or industrial use at a lower price but instead must be used for human consumption. For this reason, there is a great amount of old surplus rice on the market, keeping new crops from reaching the consumer. Yet, the memory of the government spending 3 billion yen to get rid of surplus rice during the 1970s and 1980s is still fresh in my mind.

Orito: The governmental stored rice is seemingly a kind of welfare benefit for citizens in case of food shortage as well as a security measure against rice gambling. The government should just save emergency rice without any conditions attached. From our experiences we only need to save 2 million tonnes of rice. If there is a surplus, the rice should be used for livestock feed and consumed by us in the form of pork or eggs. We are not the only ones who should eat rice.

2.2.3 Agriculture and the Environment in Japan

From a bird's eye view agriculture is the greatest friend of nature. If viewed from the ground with insects' eyes, agriculture is the greatest enemy of nature. Agriculture is deeply connected with global environmental problems such as global warming, depletion of the ozone layer and acid rain. On the other hand, farms and rice fields as well as forests possess the power to absorb the destructive forces of heavy rains, drought and typhoons and alter them into positive energy as well as the capacity to produce organic substances for water and soil resources. Green rice fields are natural air purification systems that change carbon dioxide caused by automobiles and industries into oxygen through the photosynthetic process. Acts of nature and humans are combined in terrace rice farming to provide food for wild birds and animals. (See Figure 7-6.)

Japanese annual rice production comes to 3,200 billion yen and according to calculations by the substitutive cost method, the total value of environmental externalities provided by 2,830,000 hectares of paddy fields is estimated to reach 4,700 billion yen. (See Table 7-9.) Agriculture is a part of our natural and cultural heritage and should be considered to be not only a system of food production but also a system for conserving ecological systems, land, water and landscape.

On the other hand, agriculture can be a major cause for environmental destruction. The land is cultivated into farmlands at the expense of tidelands and marshlands. In spite of the crop surpluses, forests and grasslands are bulldozed in the name of development. Natural water paths are covered with concrete for the sake of improved productivity. More immediate attention should be paid to the restoration of farm communities left abandoned rather than the creation of new farmlands. Agricultural policies must be implemented locally in order to deal with this problem.

Even seemingly beautiful terraced rice fields on a mountain can be a source of pollution. Its drainage often contains residues of nitrogenous fertilizers and other agricultural chemicals due to excessive inputs and causes water pollution. Sixteen out of twenty three harmful substances regulated under the Water Pollution Prevention Act are agricultural chemicals such as simazine, benziocarbon and diazinon. Chloronitrophine (CNP), which had been used nationwide as a herbicide for 29 years, was found carcinogenic and its use was banned by Ministry of Agriculture, Forestry and Fisheries as of March 7, 1993. Discarded vinyl used for greenhouse cultivation is another cause for environmental pollution. N_2O generated by nitrogenous fertilizers and methane (NH_4) emitted from paddy fields and manure have more impact on the greenhouse effect than carbon dioxide (CO_2) and spur on global warming.

"New directions for food, agriculture and agricultural policy" (a new policy) implemented in 1993 seeks to improve standards for agricultural chemical inputs in order to establish and sustain environmentally sound agriculture. This new governmental policy does not seem very innovative since this kind of attitude has been existent in people's minds since Meiji period. In order to meet the challenge we are facing now, the new policy on environmentally sound agriculture must seek drastic change in Japanese agricultural systems, based on more innovative concepts that are not included in the Agricultural Basic Law.

There are two directions. One is to implement Low Input Sustainable Agriculture (LISA) as proposed by the 1988 U.S. Farm Bill and compensation programs for environmentally sound agricultural practices as implemented by the EC in 1985. The other is to expand local distribution and consumption of farm products, which is often referred to as "produce locally, consume locally."

The value of economic efficiency and money that are highly regarded by industrialized societies are often not shared by agricultural communities whose lives and livelihoods are directed by the cycle of nature. Agriculture relies on the sun, soil and water, things that have an inconvertible value in environmental and cultural aspects.

In 1975 the EC implemented a policy providing farmers who live in "areas facing adverse conditions," especially mountainous areas, with income security in order to balance their income with farmers in other areas. Currently the average income compensation provided by the EC is 155,000 yen per farm. In England, where the farms are larger in area, it is 560,000 yen. On average it is 8.7% of total agricultural income. The ratios of "areas facing adverse conditions" against the total area of croplands in each EC counties are 54% in the area of the former West Germany, 45% in France and 53% in England. Since 1985 the EC has been providing compensatory funding for the loss of income to farmers that contribute to the preservation of local natural surroundings and landscape through low input farming practices.

2.3 Ecological Farming in the United States

In Willow, Glenn County, California, 300 kilometers north of San Francisco, the largest rice farming area in the United States, there is a new type of farm called the Garcia Farm where ecological farming is practiced along with organic farming, in an attempt to coexist with all living things on the farm. This is one example of the new trend in U.S. agriculture since the implementation of federal guidelines for organic farm products prescribed by the 1990 U.S. Farm Bill and may suggest the future of agriculture. The author visited this farm.

Two deer jumped out of the waving golden ears of rice and quickly disappeared into the field again. A startled rabbit hopped away. A pheasant squawked overhead. A flock of ducks and wild geese descended into a pond. A swarm of dragonflies were twinkling in the blue sky above.

"We have beavers and coyotes around here, too," said Allen Garcia, as the author stood back and took in the sight. "Deer like young milky rice. Pheasants eat locusts. Many kinds of dragonflies show that this is a paradise for bugs thanks to no pesticides." "I released 2,500 catfish," continued Garcia, pointing to a flock of water birds. "Plankton grows from the nutrients coming from the rice fields and forms the anchor of the food chain for the catfish. We are using this land for multiple purposes. The land is not just producing food but providing a place to live for wild animals and recreational activities like fishing."

Garcia's grandfather settled in this land 80 years ago. After three generations, Garcia now manages 480 hectares of paddy fields while serving as the head of a local irrigation association. He does not use any chemical fertilizers or pesticides but uses cow manure, except for 100 hectares of low-input farming area. Two hundred hectares of land within the farm consists of marshland and grasslands with groves and is reserved for ecological conservation, in an attempt to live harmoniously together with wild fowl and animals. The variety of rice he grows is Kokuho Rose, which is gaining attention even in Japan because of its reputation for high quality and high price due to its limited yield unlike other imported rice that is inexpensively mass-produced.

"Consumers are willing to pay for the cost of rice eaten by deer and birds. Agriculture is keeping us from destroying nature and polluting the environment. Rice culture is an industry that can support so many lives," said Garcia. His words have a great importance especially to farms in this region in Sacramento Valley where yield of crops is declining by 15% due to air pollution caused by automobiles.

Garcia's farming technique, however, has some weaknesses. The yield of rice is 30% less compared with other farming methods. Although the loss can be covered by the saved harvesting cost by not using agricultural chemicals, he goes one step further and leaves 200 hectares of paddy fields fallow alternatively every year in order to foster soil productivity. Rice farming in the U.S. has been producing surpluses and 20 -30 % of rice acreage has been cut down. Garcia's strategy is aimed at sustaining his farm by optimizing the governmental compensation for fallow fields. Garcia says, "Prices for farm produce are too low in the U.S. Engel's coefficient in the U.S. should be raised from the current 14% to 28 to 40% like other developing countries. Most importantly, we must realize that agricultural practices in the U.S. are becoming the worst kind ever, polluting essential natural resources like soil and water by overexploitation and high chemical inputs."

Garcia claims that consumers should be responsible for the cost of ecological farming. According to a public opinion-poll, 80% of state residents prefer organic farm products whereas only 1.5% of them are willing to pay for the added production cost, which is, on average, 33% higher than other farm produce. This is the reason why farmers like Garcia are still in the minority. Farmers like Garcia who belong to the California State Organic Farmer's Association, producers of organic farm products certified by the state, composed only 1% of the total number of rice farmers in California when the1990 U.S. Farm Bill was enacted.

At the national level, however, organic farming no longer seems to be a rare agricultural technique. Total organic food sales increased remarkably from 174 million dollars in 1980 to 1,250 million dollars in 1989, reflecting the rising public concerns over food safety and health. Organic food sections are no longer rare in many supermarkets. Twenty-three states have mandated their own regulations on organic food, although they vary from state to state. For example, under Minnesota state law products must be harvested in a farm that has been totally chemical free

Table 7-10. Sustainable Agricultural Practices in Harmony with the Environment.

Techniques	Description
Reduced use of chemical fertilizers and pesticides	* Adjusting/reducing amount of agricultural chemicals * Improving application methods (adjustable effect, local application * Using alternative methods (organic fertilizers, pheromones, natural enemies, physical control)
Organic farming	* Eliminating use of synthetic agricultural chemicals
Soil management	* Maintaining soil productivity through the use of organic substances * Maintaining soil productivity through crop rotation and cleaning crops
Joint operation of farming and stockbreeding	* Recycling farm byproducts (e.g., straws) and livestock waste through systematic or region-wide farm management. * Recycling organic waste, such as livestock feces and food remains (composting, returning to soil)
Promotion of livestock industry enhancing cyclical use of resources	* Improving management and application of livestock waste * Implementing site-specific grazing methods using abandoned farmlands and pasture/woodlands.

Source: Ryohei Kada, Environmental Protection and Sustainable Agriculture, Ie no hikari kyokai, 1990

for more than three years to be certified as organic, whereas in California the period required is only one year.

The organic farming section of the 1990 U.S. Farm Bill was proposed by Democratic Party legislators and enacted with great support of the Republicans Party. It seeks to increase consumers' awareness of organic food standards and to promote organic as well as ecological farming techniques among producers. Its goal is to nationally standardize the organic certification process and to foster interstate distribution of organic products. The 1990 Farm Bill also created a National Organic Standards Board in order to implement and monitor the national standards with the cooperation of producers, distributors, processors, researchers, consumers and government officials.

The standards are quite strict. Use of synthetically compounded fertilizers, pesticides, soil conditioners and smoke generators is prohibited. It also regulates the use of growth hormones, artificial coloring, preservatives and other additives in post harvest products for the purpose of distribution, processing and storing. In order to be labeled "organic food," a product must be produced on a farm that has practiced organic farming methods for three years before its harvest.

This national trend toward organic farming is in response to increasing public demand for food safety triggered by the environmental concerns that arose as a result of agricultural practices relying on chemical substances such as land degradation, water pollution and rapidly increased pests and pathogenic bacteria resistant to agricultural chemicals. Garcia's attempt is in line with this trend and he looks further into the future by seeking to improve consumer's quality of life by conserving the natural environment through joint efforts between producers and consumers. His effort to preserve the ecological environment through his agricultural practices is worthy of being called "environmental democracy" and rises far above any self-serving interests.

One of the reasons for the development of low input sustainable agriculture in the U.S. is increasing topsoil erosion, which has become a national concern as serious as the drug problem. According to the U.S. Department of Agriculture Soil Conservation Services (SCS), 5 tonnes of topsoil per acre (one acre equals 4000 square meters) of a third of the 400 million acres of entire croplands in the U.S. is being eroded each year. The affected areas are marked red on the map prepared by the SCS. Still overall grain yield has not decreased due to invested production cost. It costs $1.5 per foot to extend the height of a 5-ft terrace farm and $1.7 per foot to lay a drainpipe. Moderately eroded croplands require additional 4.53 kilograms of nitrogen, 800 grams of phosphate and 2.7 kilograms of potash for each acre. Soil erosion is not just a loss in agricultural economy. According to the Conservation Fund, a non-profit environmental protection organization based in Washington, D.C., soil erosion is causing many problems and the damage is $6 billion (800 billion yen) annually. Drinking water is contaminated by residues of agricultural chemicals seeping out of the soil carried into reservoirs and dams. Eroded soil clogs channels making them un-navigable for ships, and kills many fish.

The problem of topsoil erosion has an international origin. Traditional farming practice in the Midwest prairie regions of the U.S., known as the "bread basket of the world," with its severe climatic conditions has been one of crop rotations, following the usual pattern of soybeans-corn-pasture-soybeans and terrace farming on slopes. Such agricultural techniques have affected soil productivity. During the 1960s and 1970s the crops produced in the U.S. were in high global demand due to world population growth, GDP increase in leading industrialized countries and the decline in agricultural output in the former Soviet Union. Many farmers switched to continuous cropping of highly-priced crops, which eventually caused soil erosion in an attempt to meet this demand. The erosion was not more than 2.7 tonnes per acre when crop rotation and terrace farming were the major agricultural techniques, whereas currently it is 19.7 tonnes on average, and 25 tonnes in southern Iowa where the land is steeper. Since soil productivity decreases when more than 3 tonnes of soil is eroded per acre, this can be described as "erosion of the Earth." Topsoil erosion can be fatal to the farm economy since it leads to income loss. For example, loss per acre in yields by moderate erosion is 16 bushels for corn, 5 bushels for soybeans and 9 bushels for oats (1 bushel = 35 liters).

3. FREE TRADE AND ENVIRONMENTAL DESTRUCTION

The Japanese government's 1996 White Paper on the Environment points out that agriculture in developing countries concentrates on production of commercial crops for export such as coffee, pepper and pineapples and cannot sustain domestic food supply, leaving those countries no other choice but to rely on imported staple foods from developed countries. The paper says that the result is environmental destruction and pollution in each country. It also reports that a survey conducted of 250 women working on a plantation in Malaysia who sprayed lindane, a pesticide prohibited in Japan, found that the workers complained of dizziness, muscle pains, burns, nausea and nail discoloration. Although these women had been provided with protective suits, it was too hot to wear them.

In the Philippines, Indonesia and Papua New Guinea, coral reefs, underwater tropical forests fostering many marine lives, are dying because of a widely used fishing method in which fish or shrimp are caught after being chased into a coral reef and paralyzed by deadly cyanide poison. Twenty to twenty five thousand tonnes of these fresh fish are exported to Hong Kong annually, and the fish are gaining popularity among Japanese tourists. Some of them are imported to Japan for aquariums.

Japan is the largest food importer in the world. This means that only 2.2% of the global population is importing the largest amount of resources including fish, meat, timber, livestock feed, corn and soy beans.

Forty percent of kitchen waste in Japan consists of leftovers or stale food. While food resources are being wasted in Japan in this way, 800 million people in developing countries are starving. While the Japanese government's White Paper on the Environment reports the current environmental situation in South East Asia, it is perhaps an unwitting critique of the Japanese government itself since this situation was brought about by its own policies. The government should face these arising problems by recognizing the priceless value of the environment and work toward a solution.

In 1964 when restrictions on timber imports were lifted, Japan started importing timber from Thailand and then from the Philippines. Both countries quickly lost their tropical forests. Another timber exporting country, Indonesia, realized the seriousness of the matter and eventually banned log exports. Since then, Japan has been importing logs solely from Sabah and Sarawak in Malaysia. In Sabah, with the area of about the size of Hokkaido, most economically profitable trees have already been cut down. In Sarawak, in spite of Prime Minister Mahatir's repeated denials, tree felling had to be decreased by 30% for the purpose of "sustainable development" to comply with recommendations from a study mission sent by the International Tropical Timber Organization (ITTO).

What is the significance of these events happening under the system of free trade? The author learned on a visit to the village of Kabatasan, Sabah, where traditional slash-and-burn farming was practiced, that intervals for controlled burning had declined from the traditional 20 years to 6 years. This was caused by the fact that land suitable for slash-and-burn farming was less available as a result of expansion of large farms which produced palm oil for the internationally market. Although the statistical report by the Food and Agricultural Organization of the United Nations (FAO) considers the slash-and-burn farming method practiced by poor farmers as the cause for deforestation, it does not explain in detail the circumstances that forced farmers to encroach upon tropical forests.

There has been confrontation around the world between governments promoting free trade of every product from tropical rain forest timber to agricultural products and consumers who demand environmental conservation. In Europe some local governments in Germany and the Netherlands stopped using timber produced from rapidly disappearing tropical rain forests. Their initiatives are based on a criticism that more free trade will stimulate consumption of forest resources and accelerate the deforestation of tropical rain forests in developing countries suffering from cumulative debt, resulting in continued exploitation of forest resources.

At a trade conference held in Angers, France in 1991, Eiichi Nakao, then Minister of International Trade and Industry proposed four principles for trade regulations for the sake of environmental conservation. His proposal included accepting minimal import restrictions on products which are connected with obvious harm to the environment, if domestic environmental standards have been met and international consensus for restrictions exists.

For example, in the U.S. the use of 62 kinds of agricultural chemicals are approved for post-harvest storage purposes, including application on wheat of a pesticide called Marathon. When Japan imports such products, monitoring the safety level of residual chemicals is difficult. Thus Japanese consumer groups asking for safer food regulations are opposed to such imports and claim the government has sold out food safety in the name of free trade.

In California there was a local referendum on an initiative called "Big Green" in 1990. The initiative sought to ban the use of pesticides that were potentially carcinogenic or thought to cause genetic disorders as well as a ban on the sales of agricultural products if residues of such chemicals were detected. Although the initiative was voted down, it opened a debate on the free trade of agricultural products and their safety. Opposition factions criticized the initiative saying that it would prohibit free trade by creating the same kind of tariff barriers that proponents of the General Agreement on Tariffs and Trade (GATT) sought to tear down and would be tantamount to California's secession from the U.S. On the other hand, supporters argued that the free trade of agricultural products that GATT promotes did not take safety standards into consideration, thus neglecting environmental conservation.

Figure 7-8. Trends in Profitability.

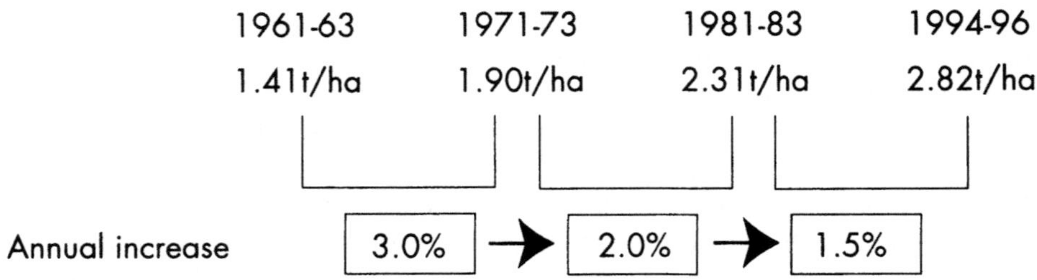

Source: FAO, FAOSTAT.

As stated earlier, however, the use of 62 kinds of post harvest chemicals is already approved in the U.S. This is one aspect of agricultural trade. Moreover, exporting countries tend to consider initiatives like "Big Green" as trade restrictions that limit free trade in the name of environmental protection.

Figure 7-9. Change in world crop production, production per unit area, harvest and arable land area.

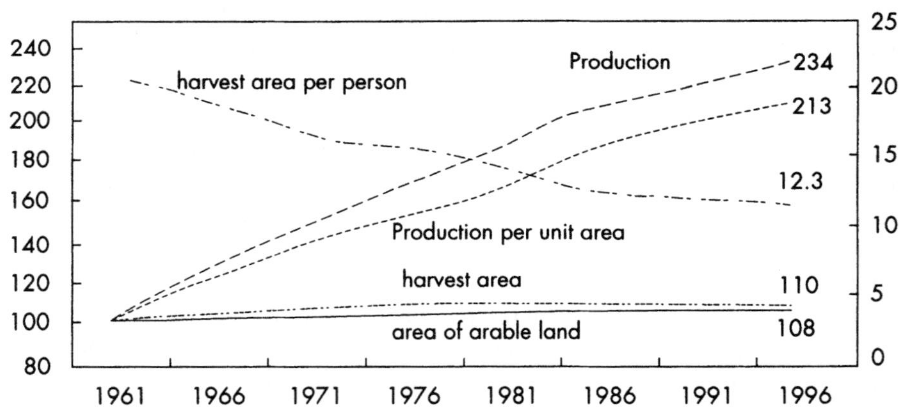

	1961-1963	1994-1996
area of arable land (10 million ha)	12.7	13.6 (1994-1995)
crop harvest area (10 million ha)	6.5	7.0
crop production (10 million tonnes)	9.2	19.7
crop production per area (tonne per ha)	1.4	2.8
population (10 million)	31 (1962)	57 (1995)

Source: FAO "FAOSTAT"

Development of the global economy is dependent on the promotion of sustainable agriculture, which will contribute to lasting natural resources and environmental conservation. As long as the root cause of global environmental destruction lies in the production and consumption systems in use by developed countries and the exploitative use of natural resources in developing countries, future GATT negotiations must also face the issue of environmental conservation if the participants are to achieve their goal of continuing economic growth. In response to the four principles proposed by Nakao, debate continued through intergovernmental discussions on the issues of trade and environment within the GATT Uruguay Round and the World Trade Organization (WTO) that succeeded it.

4. SUSTAINABLE AGRICULTURE PROVIDING A BASIS FOR THE NEW AGRICULTURAL BASIC LAW

The Investigative Council on Basic Problems Concerning Food, Agriculture and Rural Areas, an advisory committee that reports to the Prime Minister, has been working toward the formulation of the "New" Agricultural Basic Law which will suggest the basic direction of future government agricultural policies. On September 19, 1998 the Council submitted a report to Prime Minister Obuchi, focusing on (1) domestic agricultural production as the basis for a stable food supply, and (2) enabling agriculture and rural areas to fulfill their multi-functional roles. The report analyzes the future trend as described below, emphasizing the need for implementation of sustainable agricultural systems that seek to reduce the use of agricultural chemicals.

We are now at a turning point in our civilization where values and lifestyles are greatly changing. Since the beginning of this century we have been in pursuit of materialistic gains and have been to achieve economic growth through technological progress and development. Now we are concerned about the limits we have put on the Earth's resources, the environmental problems we face and the impending crises of food shortages. As we experience global crises and uncertainties, we must thoroughly review our way of life from a broader point of view. We now pursue spiritual as well as cultural affluence and tend to appreciate such values as spiritual comfort and contacts with each other equally.

In this new century, we will be more concerned about the interrelationship between humans and nature (i.e., the environment), humans and humans (i.e., internationalization) and humans and their past (i.e., history, tradition and culture). New values that give priority to harmonious coexistence, human wellbeing and comfort are expected to grow worldwide, replacing the existing values placed on progress and development that supported the technology-based civilization of the twentieth century.

Food is necessary to support the life and health of the people. We must realize the importance of agriculture and the rural communities that play a key role in food production. The cyclical functions of nature are inherent in the functions of agricultural activity. Through these functions the natural environment, soil and water, is formed and preserved, enabling sustainable agricultural production. Thus the functions of agriculture should be carried out in complete harmony with nature. In addition, organic waste, such as livestock manure and food leftovers, can be reused as resources, and enhance agriculture's natural cyclical function. The recycling of organic substances is essential to creating a sustainable society.

Farming businesses should be operated on a sustainable basis and in harmony with nature and the environment so that they can fulfill their important role in preserving the cyclical functioning of nature. It is necessary to shift to farming practices that employ good soil management as the basis and reduce the use of chemical fertilizers and other agricultural chemicals (See Table 7-10).

Based on the report, the Ministry of Agriculture, Forestry and Fisheries settled "The Basic Law on Food, Agriculture and Rural Areas (New Agricultural Basic Law)". It was passed and approved on July 12, and came to force on July 16, 1999. The agricultural administration that had been based on the Basic Agricultural Law set in 1961 is to go under drastic reform. The New Agricultural Basic Law sets "Policies for Sustainable Agricultural Development" and "Policies for Development of Rural Areas" in sections 3 and 4 of chapter 2. It states that sustainable development of rural area and agriculture shall be attempted by promoting the multiple function of, and by maintaining natural cyclic function of agriculture. It is worthy to note, in article 4 of chapter 1 it is clearly expressed "the term 'natural cyclical function of agriculture' means the function of agriculture in stimulating the biological and physical cycle in nature while being influenced strongly by the cycle". This is a new development pattern, which did not exist in the perspective under the old law.

5. SUMMARY

This paper has discussed the issue of sustainable agricultural development faced by the social and agricultural apparatus within three countries. It should be noted that there are common characteristics among the philosophies of sustainable agriculture practiced by its three proponents, Monkonbu Swaminathan, Kanji Hoshi and Allen Garcia. The common themes of their philosophies are working with the cycles of nature, harmonizing human activities and ecological systems, and promoting the joint participation of producers and consumers. When combined with the principle of international cooperation, the goal of sustainable agriculture that these three countries are pursuing will resonate well with the four principles- sound material cycle, harmonious coexistence, participation, and international cooperation- suggested by Japan's Basic Environment Law (1993) and the Basic

Environment Plan (1994) which seek to comprehensively and systematically promote Japan's policies for environmental conservation.

What is significant is that the attempts of these three advocates are gaining nationwide support from local communities to national governments. This movement does not merely represent an alternative method that can be practiced alongside existing agricultural methods. The economic success of Takahata and Garcia farms is regarded by other farmers as a model for the agricultural industry of the near future. We must realize that the abolishment of the food control system in Japan and development of the LISA program in the U.S. reflect the rising demand for reform of standardized agricultural systems in developed countries that have been emphasizing producers' economic profitability at the expense of consumers needs. We should also realize, before it is pointed out in a UNFPA report, that experimental bio-villages in India suggest that agriculture must be the primary industry in Asian developing countries, and that these countries need to develop a mechanism of stable agricultural production and distribution in order to achieve sustainable development of their societies as a whole.

NOTES AND REFERENCES

1 Excerpted from Ryohei Kada, Environmental Protection and Sustainable Agriculture, Ie no hikari kyokai, 1990

2 Source: The State of World Population 1993

3 Translator's note: *Kyosei* means coexistence or living together and *Juku* here is a old-fashioned way to say private school.

NOTES AND REFERENCES

Section II

New Development Patterns and Regionality

Chapter 8

NEW DEVELOPMENT PATTERNS IN ASIA

SHINJI FUKUKAWA
Dentsu Institute for Human Studies

1. IMPLICATIONS OF GLOBAL ENVIRONMENT PROBLEMS IN THE HISTORY OF CULTURAL PROGRESS

1.1 Civilization and Global Environmental Problems

Four point six billion years have passed since the Earth was born in space and two million years have passed since humanity appeared. It is no exaggeration to say that this planet, which has been the beautiful home and source of survival for humanity, is now threatened.

What I would like to emphasize first is that the current global environmental problems may bring about a crisis that could never have been anticipated throughout the history of civilization. Since the very dawn of history, humanity has been pursuing technological development to protect himself from and conquer nature. However, economic activities triggered by these technological developments have grown to a scale that it now threatens to destroy natural cycles.

Humanity has reached the point where it must reconsider industrial systems based on economies of scale that it pursued since the industrial revolution, technological developments seeking to improve the efficiency and functions, and lifestyles based on convenience through using disposable goods. The destruction of the global environment would signify the destruction of our civilization, which has been developing for millennia. Consequently, as we reach the point at which global environment is under threat of destruction, we must distance ourselves from the paradigm of the culture of constant growth.

K. Matsushita (ed.), Environment in the 21st Century and New Development Patterns, 201–213.

When I ponder these issues, I recall traditional Japanese gardens, which possesses harmoniously both the wonders of nature and artificial beauty. They integrate the beauty of art we create and the magnificence of nature represented by flowers, birds, breeze and the Moon. Perhaps the gardens symbolize a path, which shall restore the functions of nature: harmony between economic growth brought about by human ingenuity and the conservation of the global environment.

1.2 Population Change and the Global Environment

Global environmental issues are closely related to the problem of world population. The population of the world, which was only one hundred million at the time of the birth of Jesus Christ and about one billion around the time of the industrial revolution, rose to nearly six billion by the end of the twentieth century. It is predicted to reach around 10 billion by the middle of the twenty-first century.

This population explosion will certainly increase pollution of the global environment by increasing the consumption of resources and energy and creating mountains of waste. At the same time, a gap between the rapidly increasing populations in developing countries and the stagnant growth or declining populations in advanced countries may cause serious confrontations.

The high population growth in developing countries means that consumption of resources and energy will likely increase accordingly even if per capita incomes remain at the existing level. Consequently, it is inevitable that deforestation will increase considerably as people seek fuel for daily use, and farmland will be devastated due to the greater need for cultivation. Destruction of the environment and food shortages will be more serious in the future.

The declining birth rates together with the aging of populations in advanced countries will cast a dark shadow on their potential for economic growth in the long run. I must say that low population growth in advanced countries is very likely to constrain economic potential unless extraordinary technological innovations are made. Weakened economic growth in advanced countries would lower their burden on the global environment, but on the other hand it may also reduce their ability to transfer to developing countries the capital or technology necessary for conservation of the environment. The World Health Organization (WHO) is responsible for health and development strategies, however these strategies will not succeed unless environmental issues are also taken into account; surely they are related, as economic and environmental conditions both have health impacts.

We urgently need to solve these difficult problems that involve trade-offs between the global environment coupled with a population explosion, and economic growth, which will be accompanied by increased energy consumption.

The Earth is a single living unit where the all living things, including humanity, are mutually related in a delicate way. To solve global environmental problems, we must employ total system approaches to restore the natural balance of the Earth.

2. Dynamism in Asian Countries

2.1 Economic Potential of Asian Countries

During the 1980s and first half of the 1990s, Asia attracted the world's attention as a major center of economic growth. In fact, the rate of economic growth of East Asian countries has increased from 5.6% in the 1970s, to 6.9% in the 1980s, and 7.6% in the first half of the 1990s. If these trends continue the regional economy of East Asia, including Japan, is expected to exceed the size of the region covered under the North American Free Trade Agreement (NAFTA) and the European Union by 2010 (Japan Economic Planning Agency, 1997).

2.2 Impacts of the Asian Currency Crisis

Asian countries suffered an economic recession triggered by the currency crisis that began in mid-1997. The crisis resulted from the combination of several factors: monetary policy linking their economies to the U.S. dollar, asset inflation caused by surplus capital inflows, rise in current account deficits, fragile monetary systems, lack of market transparency, excessive production capacity and weak support industries.

In order to overcome the crisis many Asian countries were forced to undertake drastic structural reforms under the guidance and support of International Monetary Fund and aid donor countries. In reality, the currency crisis provided important lessons to Asian countries. Cynics believe that the Asian 'economic miracle' is already a thing of the past.

In my view, the economic turmoil in Asia was the result of the failure of economic policies -- in particular, monetary and financial policies -- but the fundamentals of their economies continue to be sound. Four to five years may be needed for Asian economies to get back on the road to growth. However I am firmly of the opinion that if Asian countries maintain political stability and implement the proper policies while taking the lessons from the crisis, these economies will be restored to their former dynamism.

2.3 Source of Strength of Asian Economies

In my opinion, the primary factors that promoted the dynamic growth of Asian economies can be summarized as below.

(a) High savings rates of about 30% are continuing, providing the support for high rates of investment.

(b) Loans from international and bilateral assistance institutions have strongly encouraged foreign investment and the consolidation of infrastructure.

(c) Liberalization of trade and investment has been steadily promoted in accordance with the stages of economic development, and industrialization policies have been vigorously pursued.
(d) Advanced technologies have been introduced aggressively from overseas.
(e) Highly skilled workforces have been developed.
(f) Exports to the United States and Japan have expanded strongly since the 1980s.
(g) The emergence of middle classes has expanded consumer markets in these countries.

I should say that the development of Asia led by these conditions is not simply an "Asian" phenomenon, but something caused by integration with the global economic system. This is evident from the process of their modernization and industrialization since the latter half of the twentieth century.

It must be noted that high growth potential of Asian countries can also cause great threats of negative impacts on the global environment.

3. ASIAN VALUES

3.1 Absence of Common Culture

Since the establishment of the Roman Empire, Western civilizations were only vaguely aware of Asia as a land in the "Far East" beyond the Middle East. It is interesting to note that researchers in this field unanimously agree that a unified Asian civilization or culture has never existed. Asia as a whole has never experienced a single dominant political power like that of the Roman Empire in Europe, nor possessed a common language like Latin. It has not produced a civilization having global reach like Christianity.

When I dare to search for a common sense of values in the development of Asian economies, I reach the tentative conclusion that it is "modernization" itself, since most of the Asian countries are seeking to develop modern industry, construct modern cities, and enjoy modern lifestyles. On the surface, these could probably even be termed "de-Asianization."

With regard to religion, the ultimate form of spiritual activity, Buddhism, Confucianism, Hinduism, Islam and Christianity all coexist in Asia, giving no evidence of a common cultural religion. Neither Buddhist nor Confucian civilizations became ideologies that dominated the entire of Asia.

3.2 Common Features Amidst the Diversity

Is it impossible to find even a marginally common set of values in Asia? Considering the fact that Confucianism, Buddhism and Islam have had major influence on Asian countries, several common features can be found in the people's way of thinking centered on these religions.

Pluralism

The first feature is that they admit a pluralism of values. In Asia, numerous races, languages and cultural traditions coexist, respecting each other. Consider the process of liberalization of the Asia-Pacific Economic Cooperation forum (APEC): each participating country is to voluntarily and individually implement liberalization programs based on the principles agreed at the APEC Osaka Conference in 1995. In my view, this approach symbolizes the plurality of values in Asia. I feel that Asian people have the ability to distance themselves from thoughts of domination and to accept the values and ideas of others.

Confucianism stresses the significance of "harmony." This idea can be taken as pluralism that accepts other values. Buddhism states that life is empty and vain, a view that is far from being exclusionist. It is my understanding that this pluralism is on a completely different basis from the ideas that took root in the dualism established in Europe.

Islam, which stresses the law of Allah, is a monotheistic religion just like Christianity. However, the fundamentalist character of Islam faded by the time it reached Asia. Hindus in Bali do not adhere to the caste system. Even in India, the motherland of the Hindu religion, the caste system is beginning to weaken with the rise of the middle class.

Because agriculture provided the foundations for many Asia cultures, the principals of barter are respected, implying the existence of opportunities to make choices, and that is linked to acceptance pluralistic values. This sense of values promoted the separation of religion and politics, democratic political system, freedom of religion, and a women's liberation movement in most of Asia, enabling the introduction of modern economic and corporate systems and allowing rapid industrialization.

Harmony

The second feature is that Asians value the maintenance of "harmony." I believe this is an extension of the pluralistic sense of values. Harmony is to be found first of all in the family society. Confucianism, originally, was a religion that attached importance to the continuity of life, based on a social structure that valued the family and home, and centered on love towards children and filial devotion to ancestors. That is why blood ties are especially important in China, Korea and Japan.

Such ideology in Asia has developed into the concept of valuing human ties in local society and business communities, and leads to a way of thinking in which grouping through harmonious human relationships is respected.

In Europe and the United States, based on a sense of individualism and the adherence to contracts, there is quite a different outlook on marriage compared to Asia. The management system in Asian companies is also considerably different when compared to American and European systems in which individuals with a strong competitive spirit are praised.

Among Asian people, the principle of reciprocity and the feeling that trust should be rewarded with trust still exist. Such ways of thinking link vertical as well as the horizontal relationships. In Asia, the awareness of relationships between individuals and the group extends to the larger social scale such as the corporation and the nation, constructing a kind of social network.

The same sentiment can be seen relating to harmony and coexistence with nature. In Europe, materialism based on reason and logic expands without limit, while in Asia, the sense of values tends to limit human conduct to an appropriate size while ensuring harmony with nature and the environment. As typically seen in Japanese gardens, the method of harmonizing natural beauty with artificial beauty continues to live in the Asian culture.

Discipline

The third feature is that Asians respect "self discipline." This has something in common with the way of thinking in which harmony is valued. This tendency is common in Confucianism, Buddhism and Islam, the three basic religions in Asia that each preach diligence, thrift, and abstinence, and seek self-reflection. Of course, Europeans have maintained this kind of sense and Protestantism is considered to be the guiding principle of conduct for capitalism in Europe. Similarly, diligence and thrift were also preached as the spirit for Asian merchants in the sixteenth century, when the benefits of trade were highly valued.

In Confucianism, the ideas of being polite, working hard, and returning good faith were born from respect for harmony in human relationships, which ultimately demanded self-discipline. Buddhism also includes the concept of reward in accordance with deeds or inga-ooho, stressing diligence and thrift. The law of the Zen sect, "One who does not plough one day shall not eat for one day," is a simple statement of the concept. The high propensity to save and the high quality of the workforce in modern Asia are deeply rooted in such values.

4. CONTRIBUTION OF ASIAN SENSE OF VALUES TO RESTORATION OF THE GLOBAL ENVIRONMENT

4.1 Significance of Harmony with Nature

Looking at the situation and economic development in Asia today, there appears to be a grand effort underway to harmonize the Western style of modernization with values deeply rooted in Asia. Democratic political systems, economic systems based on market economies, and modern corporate management systems, which collectively represent the essence of modern civilization, are flourishing on the soil of Asian philosophy and culture.

I would like to ponder the hypothesis that an Asian sense of values may contribute to solving global environmental problems. Most Asian countries blame developed countries for the destruction of the global environmental today as a result of their industrialization. However, whatever the cause of the problems, preservation and restoration of the global environment are important issues that all of humanity must work together to solve in the twenty-first century.

If Asian countries are to continue on the path of sustainable growth, preservation of the global environment is essential, as well as maintaining stable supplies of food and energy. Considering that half of the predicted carbon dioxide emissions, a primary cause of global warming, generated in the year 2020 will come from Asia, Asian countries must work hard to address the problem. The Asian sense of valuing coexistence with nature should thus be useful in finding solutions.

I am quite confident that the Asian values to respect harmony with nature should contribute to reforming technological developments and social systems that will create a more healthy global environment, by efficiently using energy, eliminating pollution, recycling resources and protecting biological diversity.

4.2 Information-based Society and Asian Values

Information technology, which is currently advancing rapidly, may have the potential for contributing to protection of the global environment in various ways. Networks providing information about the state of the environment, predictions of environmental conditions, and anti-pollution technologies are sure to be useful for strengthening measures to restore the environment.

Regarding whether an Asian sense of values will adapt to the information age, I would say that in general Asian countries will perform well. Networks are central in the process of becoming information-based societies and Asian societies could be described as fundamentally "network societies," as I have already pointed out.

Although information moves via the Internet, information will be most effective only between groups with a common understanding and common base of values. The Asian sense of value that respects harmony in social relationships should help utilize information systems effectively.

The sense of pluralism should also work effectively in expanding information networks. The proclivity to accept unfamiliar information may increase the significance of network.

4.3 Technological Revolution and Asian Values

Technological innovation is a key factor that will simultaneously enable environmental protection and economic growth. From this perspective, I would like to consider whether the Asian sense of values will contribute to technological innovation.

In 1995, Professor Paul Krugman of Stanford University was criticized by Asian scholars when he openly claimed that the Asian economy was similar to the Russian economy in the 1950s, since neither of them had achieved fundamental progress in productivity through technological development. I cannot agree with Professor Krugman's view since I believe that industrial technology in Asia will certainly develop with the expansion of industrial activity. However, it is worth considering whether corporations, scientists and engineers in Asia are able to make major scientific inventions and technological innovations.

It is clear that the generally great inventions are made by just a few talented people working on a given problem. Western civilization, based on an individualistic sense of values may be superior in this sense.

However, there has been rapid progress recently in the merging of technologies from different fields, and in the linkage between scientific invention and engineering processes. Technologies related to environmental protection are a typical example of the integration of different technologies. The Asian spirit of harmonization therefore can be a favorable factor. The point is whether organizations suitable for such research and development can be set up or not.

5. FRAMEWORK OF MEASURES REQUIRED TO PURSUE SUSTAINABLE DEVELOPMENT

5.1 Change of Paradigm of Industrial Technology

The first measure required to promote sustainable development is to change the paradigm of industrial technology. Among technological approaches useful for preserving the global environment, improving technology for energy conservation is one important first step.

The energy consumption per unit of GNP in Japan declined by 36% in the fifteen years after the first oil crisis. The current level in Japan corresponds to one-half that of the United States, two-thirds that of former West Germany, one-half that of the former Soviet Union and East Europe, and one-ninth that of China. Per capital energy consumption in Japan over this period has remained almost unchanged, standing at two-fifths that of the United States and one-half that of the former USSR. Japan has also achieved excellent results in controlling pollutants such as SO_x and NO_x.

In order to achieve economic growth and preservation of the global environment simultaneously, several approaches are necessary to create a new paradigm in industrial technology.

The first is to reform the structure of the energy supply. The development of nuclear energy technology including nuclear fusion, and renewable energy sources such as solar, wind and geothermal energy will provide sources of limitless energy.

Hydropower is also a promising energy source for some Asian countries. Solar power generation in deserts with their abundance of sunlight and electrical supply through large-scale global power transmission systems should be tackled as a future dream.

The second is to improve patterns of energy consumption. Drastic improvements are needed in energy efficiency of household electrical appliances, automobile engines, industrial furnaces, production process and so on. Asian countries should enact laws to rationalize energy consumption by providing energy efficiency targets and incentives to reach those targets. Innovative technologies will greatly improve energy efficiency. Examples include solar-powered cars, electric vehicles, and solar powered home appliances, which do not emit CO_2, as well as linear motor cars and ships utilizing superconductivity and new materials.

The third is to develop innovative technologies in manufacturing processes. If bio-reactors which operate at room temperature and normal pressures, and smelting reduction processes that utilize no energy or considerably less energy can be successfully developed, they would have revolutionary effects on many industrial fields.

The fourth is to develop technologies that will restore natural cycles and create a more favorable environment. They include technology to fix CO_2 utilizing photosynthesis, technology to convert it into organic materials following natural carbon cycles and a desert aqua-net project consisting of canals interconnected with large-scale saltwater lakes.

The fifth is to improve technologies for recycling resources. When steel is manufactured from steel scrap, the energy requirement is one-third of the amount required to manufacture steel from iron ore. In this case of aluminum, the energy requirement is said to be only one-thirtieth that needed to process bauxite. Technology for recycling waste paper and other materials has advanced considerably but design and development of a new type of automobile composed of easily reusable parts, development of biodegradable plastics, and other such technologies are still needed.

I expect that Asian countries will take up some of these challenges in order to reform the paradigm of industrial technology, by introducing advanced foreign technologies and raising their own capabilities.

5.2 Improvement of Life Styles

It is difficult to deny that an economic system driven by consumption, based on the concept that consumption is a virtue has brought about the economic growth the world has experienced. The pursuit of lifestyles of convenience and comfort through the widespread use of disposable goods has invigorated many economies. At the same time, it is true that through these lifestyles humanity is losing energy resources and the environment, both precious for humanity. Most large cities are disposing of excessive industrial and daily waste in volumes beyond their capacity to treat them and are causing secondary pollution by illegally dumping waste in the oceans.

Ultimately, it is necessary for us to review our lifestyles and attach more importance to the environment through self-restraint on wasteful conveniences and cooperation in the recycling of resources. For this, education and awareness-raising must play a role in lowering our dependence on energy and resources.

As discussed earlier, the sense of values in Asia may have some sentiments and elements that encourage the improvement of lifestyles in a way that people may be able to coexist with nature. Education on environmental protection, the provision of necessary information and extensive activities of non-governmental organizations are quite useful to stimulate this sense of values.

5.3 Improving Industrial Systems

I believe that the strategies of corporations and initiatives of corporate executives are of utmost importance in order to solidly establish environmental protection in industrial activities. For this purpose, the government should provide the essential information, establish reasonable standards of environmental protection, provide effective incentives or disincentives, and stimulate research and development.

Reflecting on the experiences of the 1960s and 1970s when industries in Japan made great efforts to reduce environmental pollution, industries in Asia should recognize the polluter pays principle (PPP) and internalize the cost of environmental protection into the cost of doing business or add those costs into the selling price. Fair and efficient procedures of environmental assessment must also be established. In order to make energy saving and environmental protection measures the norm in business activities, self-monitoring and self-regulation mechanisms must become integral parts of corporate management.

In Japan, legislation has made it mandatory for designated corporations to employ certified experts on efficient use of energy and environmental protection in order to inspect and improve energy consumption and anti-pollution processes. I think it is important to introduce such system elsewhere, by which market forces would stimulate the efforts of businesses to reduce environment costs through competition.

What I want to stress is that it should be the private industries that take the initiative to establish the best mix of environment protection and economic development. Corporations should be required to apply and introduce the best technologies and systems when making investments and conducting business operations. They must also undertake environmental assessment honestly and fairly.

Corporations should also heighten the transparency of their business activities by expanding dialogue and communication with local media and concerned non-governmental organizations. When choosing goods and services, consumers pay increasing attention to the environmental burden of those goods and services and to the extent to which suppliers contribute to the preservation of the environment.

The publishing of environmental reports or "clean progress reports" are certainly good ways to promote transparency. Environmental certification systems and eco-labeling are also effective tools to provide useful information to consumers.

6. COLLABORATION WITHIN APEC

6.1 Three E's to be Achieved

In order to realize sustainable growth after the currency crisis, Asian countries must strive to simultaneously achieve progress with three "E's", namely economic growth, energy supply and environmental preservation. I believe that APEC can provide a good foundation to achieve the goals by facilitating exchanges of experiences and technologies between member countries, and encouraging joint action.

6.2 Growing Energy Demand in Asia

According to International Energy Agency (IEA) predictions, the demand for primary energy in Asia will increase by 4% annually toward 2010, twice the global rate. Most Asian countries plan to expand the use of coal and nuclear energy, however, it is predicted that most of the increased energy supply will come from oil.

According to forecasts by Japan's Ministry of International Trade and Industry (MITI), Asian countries will import 680 million tonnes of oil in 2010, twice the 335 million tonnes imported in 1992. Dependence on oil from the Middle East will climb to 80%.

As Asian countries inevitably industrialize and motorize, the demand for secondary energy such as gasoline and electricity will grow considerably. The needs for financing will be enormous for investments to boost supply capacity of secondary energy sources.

6.3 Issues APEC Should Address

Considering the future energy scenarios in Asia, APEC should undertake the following tasks.

(a) Create common understanding among APEC member countries on the energy situation by making joint forecasts of energy supply and demand.

(b) Focus policies on energy conservation, and promote the transfer and development of related technologies.

(c) Emphasize diversification of energy supply sources and development of new energy.

(d) Strive for efficient transportation systems.
(e) Develop regional concentrated heating and cooling systems.
(f) Promote energy-efficient urban development.
(g) Cooperate with each other to secure funding for the energy development and environmental protection.

6.4 Development and Transfer of Technology for Environmental Protection

It is estimated that emissions of SOx and NOx in Asian countries will double by 2010, as a result of increasing energy consumption. Half of the predicted increases in global CO_2 emissions by 2010 will come from the Asia-Pacific region. In this regard, APEC should also undertake the following joint initiatives for the protection of global environment:

(a) Promote the smooth transfer of anti-pollution technologies such as for clean coal, de-sulfurization, de-nitrification and energy conservation. Countries should also cooperate to develop technologies adapted to local conditions, secure funding for anti-pollution facilities, and expand workforce training programs. The Green Aid Plan undertaken by the Japanese Government since 1992 may provide useful tools for this target.
(b) Establish databases on environmental and energy technologies to facilitate technology transfer (e.g., the APEC Virtual Center for Interchange of Environmental Technology, International Center for Environmental Technology).
(c) Jointly develop innovative technologies. The New Earth 21 project is working on the development of carbon sequestration, renewable plastics and new production processes using bio-reactors. If this work succeeds, the new technologies will help make big strides in sustainable development.
(d) Elaborate applications of the Clean Development Mechanism (CDM), Activities Implemented Jointly (AIJ), emissions trading and other types of international collaboration.
(e) Work towards the construction of eco-cities, with high energy efficiency, zero emissions, and promotion of greenery.
(f) Establish inspection systems, set appropriate standards, which create the proper incentives, and promote greater corporate and public awareness to pursue sustainable development.

7. GLOBALIZATION AND THE IMPORTANCE OF THE ENVIRONMENT

Disasters which forced us to recognize global environmental destruction have occurred around the world in recent years.

Recent flooding on the Yangtze River destroyed 56 million houses and depressed the Chinese GDP by an estimated 0.4% that year. A heat wave attacked the southern part of United States, killing more than 150 persons. The El Nino phenomenon caused serious damage to farmlands and forests in Indonesia. I believe that destruction of the environment will harm the political stability of affected countries and the foundations of the global economic system by negatively affecting living and health conditions and upsetting the stability of energy and food supplies.

Asian countries are striving to adjust their economic systems to global standards in order to recover from the currency crisis. With the knowledge that environmental destruction may threaten global security, leaders in Asia must fully recognize that the protection and restoration of the global environment are prerequisites to achieve stability of the political, economic and trading systems around the world.

Chapter 9

NEW DEVELOPMENT PATTERNS IN INDIA

HARI SRINIVAS
United Nations University

1. INTRODUCTION

India is a country of striking contrasts and immense ethnic, linguistic, and cultural diversity. There are more than 1,600 languages, nearly 400 of which are spoken by more than 200,000 people.[1] Many of the 25 states that make up India's federation are larger than most countries. Thirteen states have more than 20 million people, six have populations of 60 million, three exceed 80 million, and one has more than 140 million people. These states differ vastly in terms of their natural resources, administrative capacity, and economic performance.[2] As a whole, India covers a land area of about three million square kilometers, roughly one third that of the United States (GoI 1998).

Figure 9-1: States of India.

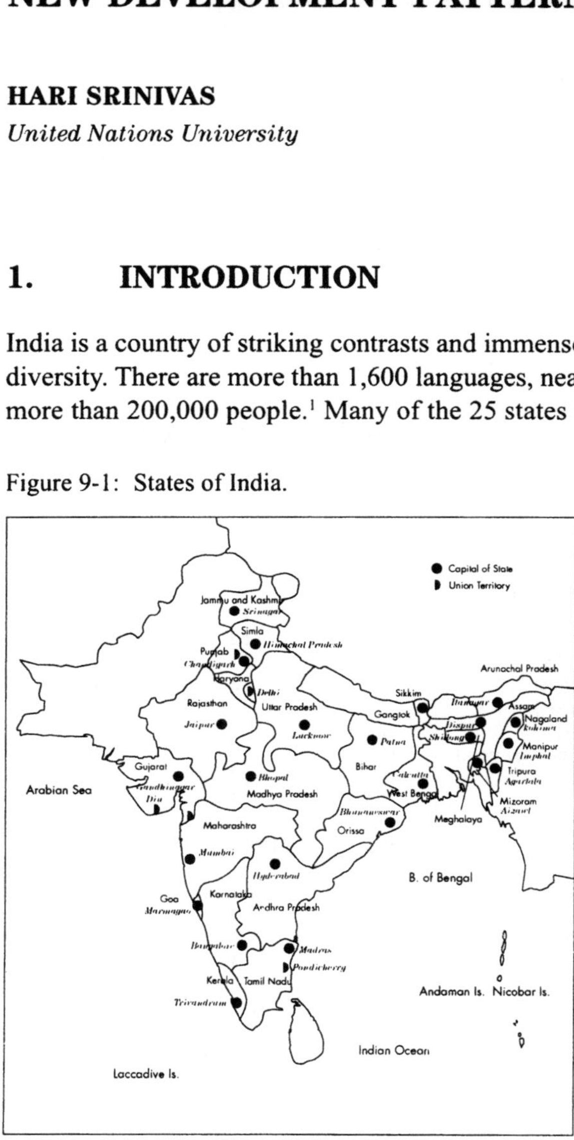

India is still largely a rural country. Agriculture plays an important but diminishing

K. Matsushita (ed.), Environment in the 21st Century and New Development Patterns, 215–237.

Table 9-1.

India: Fast Facts

- Official name: Republic of India
- Capital: New Delhi
- Government: federal republic, parliamentary system
- Estimated population, mid-1997: 1.06 billion
- Population, 1990: 846 million
- Population growth rate, average 1990-95: 1.9 percent
- Land Area: 734 million acres (297 million hectares)
- Chief crops: rice, grains, coffee, sugar cane, spices, tea, cashews, cotton, copra, coir, jute, linseed
- Industries: Textiles, food processing, steel, machinery, transportation equipment, cement, jute, mining, petroleum, power, chemicals, pharmaceuticals, and electronics
- Minerals: chromium, coal, iron, manganese, mica, salt, bauxite, gypsum, oil
- Climate: varies from tropical monsoon in the south to temperate in the north
- Environmental problems: droughts, floods, severe thunderstorms; deforestation; soil erosion; overgrazing; air and water pollution; desertification
- Languages: Hindi, English, and 14 other official languages. 200 other languages also spoken.
- Religions: Hindu (82 percent), Muslim (12 percent), Other (including Sikh, Jain, and Christian, 6 percent)
- Literacy rate: 52 percent (men, 63 percent; women, 39 percent)
- Gross national product per capita: U.S.$330

Source: WRI, 1996

Table 9-2. India at a glance.

Indicator	India	South Asia	Low-Income
Population (millions, 1997)	960.9	1,289	2,048
GNP per capita (U.S.$)	390	390	350
GNP (U.S.$)	375	502	722
Population growth (annual, %)	1.7	1.9	2.1
Poverty (% population below poverty line)	35	...	...
Urban population (% of total population)	27	27	28
Life Expectancy at birth (years)	63	62	59
Infant Mortality (per 1,000 live births)	63	71	78
Child malnutrition (% of children under 5)	...	63	61
Access to safe water (% of population)	81	77	71
Illiteracy (% of population age 15+)	48	51	47
Gross Primary enrolment (% of school-age pop.)	100	99	90
Male	110	109	100
Female	90	89	81
	1976	**1986**	**1996**
GDP (U.S.$ billions)	95.0	229.1	359.5
Gross domestic investment/GDP	20.9	23.2	25.2
Exports of goods and services/GDP	7.2	6.0	11.8
Gross domestic savings/GDP	21.7	20.4	21.9
Gross national savings/GDP	22.2	20.8	23.9
Current account balance/GDP	1.4	-2.6	-1.4
Interest payments/GDP	0.3	0.7	1.1
Total debt/GDP	15.2	21.0	26.0

Source: World Bank, 1998a

Box 1: The Balance of Payments Crisis

The proximate cause that precipitated the crisis and forced the government to undertake reforms was a severe balance of payments crisis. Foreign exchange reserves had fallen to less than three weeks of import needs. The Government ... seized the opportunity to initiate a wide-ranging restructuring of the Indian economy. The objective was to make the economy grow fast ... as a method of dealing with poverty. Rapid growth was to be achieved through better efficiency of resource use for which competition, both domestic and global, was considered essential. Private initiative was fostered through deregulation and liberalisation, which led to a remarkable transformation in a short period. Foreign exchange reserves rose (to U.S.$19 billion); exports grew at an annual 20 percent from 1993 to 1996; the economy itself grew by an annual 6 percent during the same period. Average GDP growth over the post-reforms period was 5.5 percent and inflation was brought down to around 6.5 percent.

Source: Parikh 1997

role in the economy, accounting for 28 percent of the country's gross domestic product in 1996 and 70 percent of total employment. Industry and manufacturing have expanded over the past decade, and now account for about 29 percent and 20 percent of GDP, respectively. The service sector has strengthened, growing from 36 percent in 1980 to about 43 percent in 1997. India's software sub-sector-one of the most dynamic in the world-has experienced a sustained rapid upswing, growing by 50 percent annually over the past three years.

India has made enormous strides since it achieved independence from Britain in 1947. The country's development strategy has helped it eliminate famines and bring down high illiteracy and fertility rates. India has also developed a diversified industrial base and a relatively large and sophisticated financial sector. These successes have taken place against a backdrop of India's well-established democratic system-the largest in the world-that has provided its population with an unusual degree of political freedom and stability.

Yet the development strategy that produced these results emphasised import substitution and government intervention, which ultimately increased beyond the capacity of the public sector and proved to be unsustainable. Protectionism isolated India from the rest of the world, and the country's share of world trade declined from two percent in the 1950s to less than half a percent in the late 1980s. The strategy also discouraged exports, created recurrent shortages of foreign exchange, and made the balance of payments extremely vulnerable to sudden changes in international markets. With economic growth thus impeded, poverty reduction also lagged (World Bank 1998b).

Continued economic progress is contingent upon further improvements in the human resource base. In this context, the country continues to face profound challenges in providing sufficient food, jobs, medical services, schools and infrastructure for its growing population (Ahluwalia and Little 1998). India's concentration of poverty in urban areas has significant impact in the development and spread of communicable diseases; its population is expected to surpass China's early in the twenty-first century; its industrial base-the 10th largest in the world-is a major contributor to greenhouse gases; its remarkable biological diversity is threatened by the immediate, daily needs of its expanding population (SoE 1998).

In June 1991, the country changed its course, effectively ending four decades of government-led growth. The new approach focused on stabilising the economy; reforming the financial sector, public enterprises, and the investment, trade and tax regimes; and giving the private sector a much greater role in India's development.

This economic programme produced appreciable results. Initially, growth declined sharply in response to the uncertainty of fiscal and monetary policies adopted to address the crisis. The reforms and good monsoons later helped growth rebound to five percent in the years 1992-94. During the years 1994-96 GDP grew at seven percent, placing India among the world's best performing economies.

Underpinning the economy's strong performance were important structural transformations. The declining economic role of the public sector since the start of the reform programme in 1991 is probably India's most fundamental structural change since independence. Areas that were previously the exclusive domain of the public sector-heavy manufacturing, banking, civil aviation, telecommunications, power generation and distribution, ports, and roads-have now been opened to the private sector.

At the same time, the liberalisation of the economy has also reduced distortions and increased external and internal competition. Such liberalisation measures include deregulation of insurance, urban land, agricultural exports, and foreign exchange, as well as reduction of red tape for foreign investors.[3]

Unlike previous instances of high economic growth, the recent expansion has been driven by private investment-both domestic and foreign sources. Such investment has reached an historically high level and accounts for a major portion of total investment in the country. In addition, India has begun to attract foreign investment. At $3.2 billion in 1997-98, foreign direct investment is nearly 25 times higher than it was before the economy was liberalised.[4]

In the years 1997-98, the Indian economy continued to exhibit considerable underlying strength, although GDP growth fell to 5.1 percent due to a 1.5 percent decline in agricultural production and slowing exports and industrial growth. However, pressure on the exchange rate-brought on by the turmoil in international markets, international sanctions on India following its nuclear tests, and the passage of a budget that did not contain measures to reduce the still-high fiscal deficit-led to a drop in portfolio investment (which had held steady at around U.S.$2-3 billion

Box 2: Food Production and Agriculture in India

Producing enough food for an increasing population is a critical development challenge for India. With nearly as much cropland as the United States, India must feed four times as many people. Roughly half of India's land is used for agriculture, far higher than the world average of 11 percent. Because not much more land is suitable for farming, India must develop productivity on existing lands.

Over the past four decades, India more than tripled food production. The development and rapid spread of high-yield varieties of wheat and rice in the mid-1960s was the key ingredient in this dramatic increase. Although 30 percent of this production growth (known as the Green Revolution) came from expanding farmland area, most of the increase-indeed, nearly all since the late 1970s-has come from increased average yields. In other words, farmers using the new varieties of grain were growing more food on the same amount of land.

Growth in milk and egg production also increased dramatically. Dairy and poultry enterprises now provide substantially higher financial returns per hectare than crop farming. Such enterprises create new jobs and bring about health and nutritional improvements.

Source: WRI 1996

in 1996-97) to about $1.6 billion in 1997-98. Due to the slowing pace of trade reforms, India's export growth in 1997-98 fell below world export growth for the first time in six years.[5]

2. DEVELOPMENT CHALLENGES FOR INDIA

2.1 Poverty Alleviation and Improving Quality of Life

Improving the living standards of the poor has long been among India's most important policy priorities and most pressing challenges. In the early 1950s, nearly half of India's population was living in poverty. Since then, poverty has been declining; but this has occurred slowly and vast disparities persist between and within India's states. With a gross national product (GNP) per capita of $390 million in 1997, India continues to have the highest concentration of poverty of any country, with roughly 300 million people (one-third of the population) living below the national poverty line[6] (World Bank 1998a).

Poverty remains the central issue and the biggest challenge. While the annual per capita income has been slowly rising over time, about 40 percent of the people are below what is considered the poverty line. The challenge is therefore not only

Box 3: Impact of Population growth

Rapid population growth creates many pressures. In the last 40 years, India's ecological systems have been severely taxed as the population has more than doubled, increasing consumption of resources. Poverty is a serious problem, and roughly four out of five of India's poor live in rural areas. Many of these have migrated in search of economic opportunity. When they move to urban areas, pollution levels rise, and it is difficult for cities to provide jobs, education, sanitation, and health services to ever-larger populations. When people migrate to wilderness areas, wildlife habitat is degraded or destroyed.

Source: India News Digest 1997

to find development paths and models that are environmentally sustainable but that are also socially just and culturally acceptable. For a very large number of the poor who depend on the resources of their immediate environment, poverty has been compounded by environmental degradation. Not only does the natural resource base of the immediate environment provide their basic requirements, it is also the basis of their livelihood and security. The twin challenges of poverty and the environmental problems must be taken together, as two facets of the same challenge.[7]

2.2 Providing Better Health Services

From a broad perspective, since the 1970s, the health and well-being of India's population has improved. The average life expectancy at birth has increased from 50 years to 65 years in 1998; the infant mortality rate has fallen by half to about 65 per thousand live births; and the birthrate has fallen from 6 to 3.1 children per woman. But India's social indicators continue to place it near the bottom of the ladder in most measures of human development. The United Nations Development Programme's Human Development Index puts India in the bottom one third of the low-income countries.

India's National Family Welfare Programme has helped bring the country about two-thirds of the way toward its goal of replacement-level fertility (2.1 births per woman). However, population growth and the impending strain on the environment, natural resources, and social services it will bring still pose a threat to India's development. With a total population estimated at 945.1 million people in mid-1996 (second only to China's) India remains first in the world in terms of the number added to its population each year-about 16 million (UNDP 1997).

Malnutrition also continues to constrain India's development. More than half of India's children are undernourished, and this affects their physical and mental development. Despite some improvement, India's women remain significantly more malnourished than men. Bias against women and girls is reflected in the demographic ratio of 929 females for every 1,000 males. Unlike many other countries, more women than men die before the age of 35 in India. At 437 deaths per 100,000 live births in 1996, India's maternal mortality rates remain high (particularly in rural areas), and account for almost 25 percent of the world's childbirth-related deaths.

Although declining, largely preventable diseases such as leprosy, tuberculosis, cataract blindness, and malaria continue to account for 50 percent of reported illness, and around 470 deaths per 100,000. HIV/AIDS is a newly emerging threat to India's public health and it is estimated that about three million people in India may be affected (GoI 1998).

2.3 Education as a Basic Need

Having steadily raised primary enrolment rates since independence, today India has the world's second largest education system after China, with 67 million children aged 6-10 attending primary school. However, about 32 million primary school-age children, mostly girls or those from poorer households and disadvantaged groups, are not in school; more than half of rural students drop out before completing the primary cycle, and only one-third of females make it to the secondary level.

In absolute numbers, India has the largest number of scientists and engineers in the world, but in contrast as a whole faces high levels of illiteracy and low learning achievement. Nearly half the population over 15 years old and about 60 percent of all women over 15 years old are illiterate (World Bank 1998).

2.4 Importance of Budgetary Reform

India's central and state budgetary imbalances present a serious challenge to the country's further development. These imbalances pose a major risk to macroeconomic stability, absorb funding from the private sectors, and hinder financial sector reform.

Since the major adjustment of the years 1991-93, further adjustments have been limited, leaving the central government's deficit at about six percent of GDP. At the state level, the lack of deficit reduction has put India's states in an increasingly unsustainable position (Ahluwalia and Little 1998).

The deterioration in the states' finances has been accompanied by a decline in expenditures in such critical areas as health, education, roads, and irrigation. Concurrently, there has been a continued increase in spending on loss-making state owned enterprises. Massive subsidies for power, water, irrigation, and transport have also continued.

Major reasons for the continued large central and state budget deficits are the narrow tax base and heavy subsidies that often encourage inefficiencies. Also, public sector production remains large in areas where private sector participation and production would have been more efficient (World Bank 1998b). Some states, such as Andhra Pradesh, Haryana, and Orissa, have begun to address these problems on their own, but concerted nation-wide actions, either at the national or state-level are still missing.

2.5 Financing Infrastructure

Small towns in India are showing an increasing trend towards specialisation in economic activities. Tiruppur in South India (population 300,000) generates exports in cotton amounting to Rs.30 billion; Moradabad (population 400,000) exports handicrafts worth Rs12 billion; the ancient city of Panipat (population 200,000) produces carpets and other products worth Rs4.5 billion and Biwandi (population 400,000) near Mumbai boasts more that 25 percent of the country's power looms. These and many other such small and medium sized cities are seen as a tribute to local enterprise.[8]

Box 4: Electricity production

India suffers from energy shortages and now produces about 10 percent less electricity than is needed. Unscheduled brownouts and blackouts are common, and utilities schedule power outages. Shortages have grown worse because low energy prices don't provide sufficient return to producers. Inefficiency is widespread; more than 20 percent of electricity is lost during transmission and distribution. Industries use energy inefficiently due to outdated, inefficient equipment and processes. For example, it takes twice as much energy to make steel in India as in industrialised countries. Government subsidies tend to discourage conservation. For example, low energy prices encourage farmers to overuse electric irrigation pumps, driving up agricultural demand for electricity by 15 percent annually and contributing to system overloads.

Source: ORF 1998

Besides being focal points for the regional economies, they have seen high growth in employment, much of it self-made. The regional economies are usually dominated by one or two industries. In spite of such impressive statistics, a fundamental factor that is stifling the growth of these small towns is the lack of basic networks and social infrastructure.

The towns are fighting back. Business associations are teaming up with the local government to provide piped water, education and health. Clearly, this is not enough, and local governments should foster an economic environment where local industries can have a greater role in providing basic services to inhabitants. Coordination among different government departments along with the business sector can serve a variety of economic and social goals.

Overall, improving basic infrastructure services and encouraging greater private sector participation in telecommunications, electricity, transport, and water supply can make a major contribution to growth and has been a major focus of Indian economic policy since the early 1990s. However, the injection of private capital in key infrastructure sub-sectors that for decades have been under public sector monopoly has been slower than anticipated, and so have results. Improved regulatory frameworks, and independent and empowered regulatory institutions would stimulate efficient private sector participation.

2.6 Electrical Power Generation

India's economic growth will be hampered as long as its electrical power generation is unable to meet the demand of industries and other sectors of the economy. Power shortages are estimated at about 10 percent of total electrical energy and 20 percent of peak capacity requirements. National surveys of industrialists consistently rate an inadequate power supply as one of their most critical problems (Leo's Notes 1997).

Furthermore, financial losses of the power sector continue to place a substantial burden on public sector finances. The financial performance of the sector as a whole is poor.

With responsibility for the electrical supply shared constitutionally between the central government and the states, the government now recognises the need to improve the efficiency of supply, consumption, and pricing of electricity. This can be achieved only by reforming power sector management and financing at the state level (World Bank 1998b).

The states' severe power sector problems are due to the poor operational efficiency of the state electricity boards, which are a key element of India's power system. Several other factors compound this situation. The installed capacities of power generating units have not been fully utilised, with some running at 60 percent of their capacity. Transmission and distribution losses in power networks, lack of energy conservation policies and ineffective demand management strategies are some of the other problems.[9]

A few states (particularly Orissa, Haryana, Andhra Pradesh) have initiated comprehensive power sector reform programmes such as setting up of electricity regulatory commissions with greater independence and autonomy in forming policies and tariff bases. More recently, legislation was passed to allow private investment in power transmission.

Box 5: The Chipko Movement

The Chipko movement in India is a nonviolent effort that was organized in the 1970s by local villagers in the Himalayan regions to regain control of the forests and to protect the mountain ecosystems from commercial exploitation. The local villagers tried to convince government officials and forest contractors that removing native trees even if they were replaced with commercially valuable pine trees was causing ecological instability in the region and depriving local women of their ability to find fuelwood, raise livestock, and feed their families. Locally available varieties of plants were being discarded and locally available knowledge of nature was being ignored. The people were convinced that the destructive cycle of floods and drought and resulting soil erosion that they were experiencing were directly related to the removal of the forests. Mining activities were also destroying forests and local water supplies.

Not knowing how to protect their trees from being cut down, the people banded together and literally embraced the trees to stop the bulldozers. (Chipko means to hug.) Local villagers also tied sacred threads to the trees as a token of their vow to protect these resources. Forest officials addressing villagers at a blockade in Adwani in 1977 demanded, "Do you know what forests bear? They produce profit and resin and timber." The protesters responded with a chant that has become a slogan of the movement. "What do the forests bear? Soil, water, and pure air!"

Source: WRI 1996

2.7 Agricultural Capacity and Development

Improving the efficiency of India's agriculture is key to attaining high growth and reducing poverty for the more than 300 million poor who live and work in rural areas. The growth of the agricultural and industrial sectors have been impressive as the output of food grains nearly quadrupled to 198.17 million tonnes in the 1996-97 season from 50.8 million tonnes in the 1950-51 season. In spite of this, while India spends twice as much on agriculture as its East Asian neighbours, it generates far less growth. Productivity could be improved through growth-enhancing agricultural reforms,[10] a comprehensive deregulation of domestic agricultural trade and agro-processing, and rural credit system reforms.

2.8 Managing the Environment[11]

India is one of the 12 'megadiversity' countries in the world that collectively account for 60 to 70 percent of the world's biodiversity. It is home to about 317 species of mammals, 969 of birds, and 16,000 plants. Of these, about 39 species of mammals, 72 species of birds, and 1,331 species of plants are threatened. In response, 75 national parks and 419 sanctuaries covering four percent of its land area have been established,[12] significant achievements for a nation with India's population, land-use pressures, and large mammals.

One of the most visible and established conservation projects in India, Project Tiger, chose a species on the top of the natural food chain, and effectively incorporated wide ranging efforts to conserve food chains, related species, and associated habitats and biota (Kumar, n.d.).

India's forests, grasslands, wetlands, and marine ecosystems face many pressures, including air and water pollution; cattle grazing; deforestation; over-exploitation of non-timber forest products; uncontrolled fires; and overly intensive agricultural, infrastructure, industrial, and commercial development. High population density and growth rates, coupled with poverty, accelerate this process of degradation. India's cities are beset by environmental hazards and sanitation problems due to years of under-investment and inadequate budgets, combined with population growth at twice the average national rate.

Box 6: Emerging Priorities

As India entered a turbulent election year, there were reassuring signs that the world's largest democracy reached consensus on the need for economic liberalisation. Reforms initiated in 1991 to open the economy, liberalise the market and provide a greater role for the private sector have begun to achieve results. Trade and foreign investment are up, inflation is in check, and during the past year, economic output grew by more than six percent nation-wide. India's extensive natural resources, well developed industrial base, diversified agriculture sector, and burgeoning middle class-now more than 100 million-offer the potential for rapid and broad-based economic growth that can rival its east Asian neighbours. Such growth would provide the prospect of alleviating India's chronic and massive poverty.

To complete the successful transition to a strong market economy and sustained economic growth poses real challenges. Years of socialist, inward-looking policies left a legacy of crippling bureaucracy, distorted markets, and limited entrepreneurship and innovation. Dramatic population growth-India's population more than doubled in the last 40 years-burdened urban infrastructure and threatened natural resources. The result is the world's greatest concentration of poor people, a per capita income of $300, and over 400 million people-nearly half the world's extremely poor-living below the poverty line.

Source: USAID 1997

During the 1970s and 1980s, the government established a social forestry programme that encouraged tree planting on private farms and community wood lots. Under this approach, non-governmental organizations (CSE 1998) help local communities form partnerships with government forest departments. Together they make decisions about restricting grazing in the forests, sharing non-wood forest products such as fruits, limiting tree-cutting, etc. Involving women in management groups is crucial, because women are usually responsible for collecting and processing fuelwood, fodder, food, and water. Non-governmental organizations help to promote participation by women and groups that are not part of the local village power structure (World Bank 1998b).

The difficulty of participatory forest management is in dealing with the political complexity of Indian villages, with their unique cultural traditions, various ethnicity, and division of labour between genders and among subgroups. Experience shows that for the success of joint management, decisions should be taken at the community level and must grow out of dialogue, with all village subgroups agreeing to cooperate.

Throughout India, over 1.2 million acres (500,000 hectares) of reserve and protected forestland are already under community protection through joint management agreements. This approach seems not only to be workable; it is also much less expensive than establishing tree plantations in areas that have been deforested.

3. CURRENT DEVELOPMENT PATTERNS

In a sweeping generalisation, the noted jurist, Nani Palkhivala[13], enumerates six 'fatal mistakes' of India's development process that are behind many of the ills of the 50 years after independence from Britain in 1947:

First, the greatest mistake of the initial developmental processes was to start the democracy process with adult franchise, without educating the people to be worthy citizens and to discharge their duties as citizens.

Second, to let the population increase nearly three times without any sensible or sound family planning measures and policies. This means that any economic gains achieved today will be negated by unbridled growth.

Third, was not to give sufficient importance to education in the face of increasing population numbers. This has led to the current situation where a little less than half of the population are illiterate. Education, from primary level upward, can only be a long-term programme for solving the problems of India.

Fourth, the governments have failed to inculcate a sense of pride and belonging with the diverse and ancient culture and heritage of India-as a consequence, the more than 2,000 years of history is being overlooked in the move to the next century.

Fifth, related to the one above, was not to develop a sense of national identity-Indians face the nagging question: Is India a collection of communities or is it a nation? Or is India a state without a nation?

Sixth, the de-linking of freedom (from colonialism) from a sense of duty and a sense of national direction that goes with it.

The pessimism of Nani Palkivala's six 'mistakes' has given way to guarded optimism of other leaders, whose views were presented in a recent newspaper survey[14] and illustrates the need for a globalisation of the economy that is firmly grounded in the socio-economic development of its citizens.

The growing realisation of the size and scope of India's economy has to be prioritised with compulsory primary education, particularly that of women. This is a forerunner to inculcation of the scientific temper and growth of science and technology in broader terms (I.K. Gujral, former Prime Minister).

Access to basic services, including shelter, water, sanitation and transport, has improved dramatically, but much needs to be done within the overall umbrella of poverty alleviation. Greater strides have to be taken with respect to strengthening the political democracy, focussed higher education and building up of infrastructure that matches the growth patterns of the economy recently (A.B. Vajpayee, current Prime Minister).

Politically we are seeing significant progress in a resurgence of regional alliances that have achieved a greater voice at the national level, and hence greater devolution and transfer of power and resources to the state governments and further to local governments (V.P. Singh, former Prime Minister).

Views expressed by other leaders in the survey covered immediate concerns of fighting corruption at all levels of the government, participatory development, decentralisation and self-government, youth development and employment creation that were to be tempered with long-term 'up-stream' issues such as land-ownership, gender equality and religious harmony.

3.1 Reforms at the State-level

As the focus of the 1991 reforms shifted to the states, international attention shifted to states such as Maharastra, Karnataka and Andra Pradesh that have embarked on a comprehensive programme of economic reforms on their own. These also include some of the poorer states with the worst social conditions. Project- and sector-specific investment trends at the national level are being replaced by broader state-level differences based on the investment climate fostered by different state governments.[15] Many of the states are also taking advantage of increasing state autonomy to float funds on the international market to finance infrastructure projects.

3.2 Priority on Rural Development

An emerging pattern in rural development is the reduction of subsidies on resources such as water and power; promotion of sustainable management of water, forests, and other natural resources; and improving public spending on agriculture technology dissemination and extension, rural infrastructure, and social safety nets. This is being followed up by increasing responsibility and autonomy being placed in the hands of the village community for natural resource management. At the central level, policy dialogue is being carried out in reform areas such as food, credit and fertiliser policies, and centrally sponsored schemes that are critical to the performance of the rural economy.

Box 7: Energy Scenario in India

India is expected to become more and more important as a global energy consumer. Energy demand will continue to be influenced by population growth, by industrial expansion, and by increased residential energy use by middle and upper income groups, who are using more consumer products (televisions, refrigerators and other electric products).

Energy demand could increase fourfold by 2025; carbon emissions could increase sixfold as fossil fuels replace biomass (fuelwood, crop residues). India currently accounts for 3 percent of global carbon dioxide emissions.

India is encouraging further use of renewable sources of energy (wind, solar, hydro, or biomass) particularly to solve its rural energy problems. India may be able to burn agricultural waste such as rice husks or sugar cane in conventional boilers to increase rural electrical supplies. India also has the largest solar cooking programme in the world with over 238,000 solar cookers sold at the end of 1992. Wind turbines and small-scale hydropower are also promising options for India.

Source: WRI 1996

3.3 Targeting Poverty

Poverty exists everywhere, but one factor that distinguishes poverty in India is its visibility-different estimates put 29 to 40 percent of the population below the poverty line.[16] Although there are several different dimensions to poverty, such as educational poverty (over two in five females above the age of seven are illiterate) nutritional poverty (three out of every five children are underweight and malnourished), 'expenditure and consumption' poverty is used to derive official policy directions (Dubey and Gangopadhyay 1998).

The good news is that, whatever the percentage of population below the poverty line, between the years 1987-88 and 1993-94, poverty levels declined at about 1.25 percent per year, double the decline in the 1970s and 1980s. Recent analysis has shown that rural poverty has been declining at about the same rate as urban poverty, and average real wages for unskilled and skilled farm labour have also increased. Overall, the number of people defined as poor has declined. The economic reforms of 1991-92, popularly thought to have worsened and increased poverty, have in fact reduced the number of absolute poor and improved their living conditions. This could hopefully lead to more reforms and pro-poor reforms (Ibid.).

3.4 Changing Energy Patterns

The energy sector has seen a shift in focus from centrally managed institutions towards supporting reforms at the state level to restructure their power systems and policies, including tariff reform, incorporation of public utilities, privatisation of power generation and distribution, and independent regulation. State-level initiatives in the energy sector have tended towards long-term reform in improving power transmission and the distribution of electrical power to remote and rural areas.

Coal is India's least-cost source of energy and meets two-thirds of the country's energy needs. Efforts can be observed to incorporate environmental issues into the production and distribution of coal to different parts of the country, and move towards more sustainable alternative forms of electricity generation (solar photovoltaics, mini-hydropower, and windpower).

Consumption of crude oil has been increasing at an unprecedented rate. In the years 1996-97, consumption of petroleum products was around 77 million tonnes, as compared to 30.9 million tonnes in 1980-81 and 3.3 million tonnes in 1950-51 (Leo Notes 1997).

3.5 Infrastructure Provision

It is ironic that in India, which is now self-sufficient in food production, food security issues are dominated by problems with adequacy and timeliness of distribution and access to food. One of the key approaches being used to overcome these hurdles is improving the rail and highway transportation networks that link the food-growing regions to areas that face frequent drought. Widening roads and improving road maintenance, vehicular efficiency and appropriate institutional capacity building to manage rail and road traffic are being instituted.

A comprehensive programme is transforming the face of urban India by improving basic urban services such as safe and adequate water, urban public transport, electricity distribution, telecommunications, and other network and social infrastructure, though it is clearly insufficient.

3.6 Human Resource Development

Developing adequate and appropriate human resources for the right purpose at the right level has been a critical target of human resource development policies, underpinned by a clear emphasis on primary education being made available to all children. Interrelated issues of human resource development and health, primary and vocational education and training, women and children, and job creation are being combined to increase effectiveness. A greater sensitivity for gender issues can also be seen in government policies and private sector programmes that are aimed at achieving a range of goals in family planning, nutrition, health, technical education, and vocational training.

3.7 Increasing Attention to Social and Environmental Concerns

Concern and commitment to environmental and ecological issues have been growing in all sectors in India. A vigilant civil society, along with non-governmental organizations and a vibrant media, have highlighted the negative and unsustainable effects of large development projects.[17] Emphasis on smaller sustainable projects with low impacts is being given priority, particularly at the state level.

An increased role for the civil society, as monitors and partners, has become even more critical after the recent economic liberalisation. Increased economic activity and investment opportunities have provided excellent opportunities to incorporate environmental concerns. This focus has been aimed at both the pre- and post-production processes.

Concern for broader intervention and integration of issues such as forestry and biodiversity protection, soil and water conservation, industrial pollution and prevention, and sewage disposal and sanitation have been called for, coupled with revamping of environmental policies, and monitoring and enforcing of environmental legislation.

3.8 Private Sector Participation

Many of the new programmes in agriculture, infrastructure, energy, and social sectors have come to rely on the private sector for the delivery of goods and services. Despite increased efficiency and outlays in investment, the delivery of services and infrastructure by the public sector has simply not kept pace with the demand. Hence, in contrast to the government as a 'provider' of services, there has been a move for the government to create and foster a business environment where partnerships between the public and private sectors can be built, in delivering basic services to the people. Examples of such public-private partnerships have covered feasibility studies, manufacturing, supply, building, financing, or maintaining service facilities.[18]

4. EMERGING PATTERNS

India is going through three fundamental transitions: economic liberalisation, political decentralisation, and social empowerment (Ramesh 1998). The devolution of power from national parties with decision-making power resting at the centre to more power at the state level, and from the states to the local governments, is the key trend in the political dimension. An inward-looking, state-controlled, public sector-dominated economy's transition to an outward looking, market-oriented and private investment-influenced economy constitutes the economic dimension. The empowerment of traditionally disadvantaged and discriminated groups through a variety of policies and programmes constitutes the social transition.

The transitions are not independent. Each has influenced and has been influenced by the other two. Economic reforms have been facilitated by social changes that have also transformed the political situation.

A recent analysis presented by the magazine India Today[19] lists some of the 'pro-change' dynamics that have driven these transitions:

(a) A demographic shift-two out of every three Indians are under 40 years. This means a large workforce has 'come of age' in the growth periods of the 1980s and 1990s.
(b) Growing urbanisation that has also increased economic opportunities and jobs.
(c) Spread of agricultural prosperity to new regions and multiplier effects of farm growth on other regions.
(d) Rapid proliferation of communication technology, including radio, telephone and the Internet[20]

This has given rise to a new slogan that is driving the reforms-"Globalisation of the economy, Regionalization of the politics, and Localisation of the society" (Ramesh 1998). What then are the emerging patterns of development that will drive India in the twenty-first century? This report on India has illustrated the development processes that have taken place in India over the last 50 years. It has not been a smooth or continuous process. Progress in one sector has been negated by regression in another sector within the same year; or progress in one sector has given way to a slump later on in the development process. 'Progress' in the Indian context has been uneven, with contrasting spectacular outcomes (software development, space technology, heavy industry) being placed side by side with scenarios that can be shocking (poverty, homelessness).

It is clear that there is no single development model that will work for India's socio-economic and cultural mix. A 'melting pot' of approaches, plans and strategies that stresses the one-size-fits-all philosophy probably has to give way to a 'salad bowl' of different philosophies that can maintain their individual strengths and methodologies, and yet combine within the salad bowl to generate a unique set of models and solutions.

4.1 Economic Growth

There is a clear need to focus development efforts in order to have the greatest impact-primarily through policy reform or by targeting sectors that have significant multiplier effects. Development efforts should go hand in hand with increased transparency and administrative efficiency that is necessary to attract domestic and foreign capital. A programme of continued growth of the market through policy and regulatory reforms and the introduction of new methodologies required to protect investments (Ahluwalia and Little 1998). For example, the Securities and Exchange Board of India has identified key measures in the growth of the financial markets: (a) disclosure standards for the issuers of equity and debt, (b) market surveillance, inspection and enforcement plans that meet international standards, and (c) regulations for an efficient securities trade settlement system (SEBI 1997). Inadequate urban infrastructure, such as water supply, sewage and waste treatment systems, is a significant impediment to sustained growth. It is now clear that the resources required to meet the demand for these urban services cannot come from the public sector. There is a need, therefore, to support the development of debt markets-in the form of loan guarantees, technical assistance and training-to finance infrastructure projects in India, along with increased opportunities for public-private partnerships.

4.2 Stemming Population Growth

While some states such as Tamil Nadu and Kerala have achieved zero population growth, other states have continued to have high growth rates. Strategies to stem population should aim to increase the quality, demand, and access to a broad range of reproductive health and family planning services and to address related fertility issues, including the status of women. Education of Indian women should remain the core of any strategy-World Bank studies have clearly shown that for every year of education that women receive, there is a 10 percent reduction in their fertility rates-this is illustrated by the state of Kerala, which has the highest literacy rates and also the lowest fertility rates for India (UNDP 1997 and World Bank 1998b).

Besides enhancing government-run programmes in family planning and health services,[21] emphasis should be placed on supporting the initiatives of non-governmental organizations involved in a range of health and family planning services. Upgrading of primary health care centres, supporting birth control methods, educating parents on birth spacing, and increasing access to services through the private sector are some of the measures that should be taken.

4.3 Ensuring Adequate Access to Food

With eight successive good monsoons, India's food grain production has increased markedly, but massive poverty, inadequate infrastructure and policy failures at both state and federal levels mean more than half of India's young children are malnourished and maternal malnutrition is widespread. One-third of India's population suffers food insecurity.[22] Long-term food security is challenged by India's rapid population growth (18 million people every year) and related issues of agro-technology, environmental and land degradation and climate variation.

There is a need to develop a clear strategy for the enhancement of food technology, including storage, processing, packaging, transportation and distribution. This has to be coupled with efforts to improve maternal and child nutrition and health, thereby reducing mortality rates, influencing fertility, and complementing efforts in population and health. Such efforts should work through local organizations to reach children and pregnant or lactating mothers.

4.4 Environmental Action

The combination of accelerated economic development and rapid population growth could precipitate an ecological crisis that would reverse India's recent economic gains and increase negative impacts on the global environment. Already the second fastest growing producer of greenhouse gases in the world, India could triple its emissions between 1987 and 2010 if there is no change in current practices. Carbon dioxide emissions are expected to double during the same period, and chlorofluorocarbon (CFC) emissions, which cause ozone depletion, are growing rapidly (UNFCCC 1997).

India's demand for power is expected to double in the next five years. Power plants fuelled by high ash coal are a major source of pollution in India. Clearly, policies should be directed towards increasing the percentage of power generated by 'clean technologies' such as innovative clean coal and renewable energy technologies. Air pollution control measures and technical assistance on policies and regulations also should be integrated with overall environmental policies.

India's rapidly growing cities pose crucial environmental challenges: poorly managed urbanisation, weak local governments, and the poor state of urban infrastructure contribute to deplorable conditions affecting growing numbers of families. This situation leads to environmental degradation downstream from cities that lack suitable facilities to treat, recycle or dispose of municipal solid waste and sewage. This calls for strengthening the management capability of state and local governments, community groups and NGOs with tools such as environment workbooks and risk assessments that create an accessible information base.

India is one of the foremost hosts of the world's biodiversity and loss of genetic diversity threatens many of these species; deforestation and water pollution also threaten animal and plant diversity. Research and action in this field should focus on initiatives such as genetic banks to protect germplasm, involving identification, collection and storage for long-term preservation and future access.

5. SUMMARY

This chapter presented a broad introduction to development patterns in India and evaluated the current situation. It then drew implications for the emerging patterns that India could adopt for the twenty-first century.

The emerging development patterns in India should be seen in the perspective of the recent far-reaching changes initiated by the Government of India in the years 1991-92. Faced with a severe fiscal and foreign exchange reserve crisis, the Central Government in New Delhi approached the World Bank/International Monetary Fund for loans to overcome the immediate requirements of the crisis. But with a rare long-term perspective, it also initiated far-reaching economic policies that were to result in the opening up of the economy to international participation and investment. Until then, the socialist/populist policies of the government were aimed at building and protecting local products and industries, import substitution and large state-controlled enterprises, at the cost of global competition and participation.

Liberalization of the economy initiated during the years 1992-93 is essentially characterized by easing of terms and conditions that earlier prevented foreign participation (for example, technology transfer, foreign exchange repatriation, etc.), equity participation, simplified taxation structure for both imports and exports, and a programme of de-licensing which reduced or eliminated permits and licenses required for foreign participation. The changes affected virtually every sector of the economy: trade, industry, foreign investment, finance, taxation, and the public sector. The results of these policies are clear to see: fund mobilization in international capital markets has received enthusiastic approval and support, and a jump in collaboration and joint ventures with Indian companies in providing goods and services that were earlier the unique purview of government agencies and departments.

After describing the economic, social and political conditions affecting India's development, this chapter covered the emerging development issues, in particular promoting economic growth, stemming population growth, ensuring an adequate food supply, and protecting the environment.

ENDNOTES

1 English remains a key language of use for official and governmental communication, along with the other national language, Hindi.

2 Detailed state level statistical information can be found on the Internet at http://SunSITE.sut.ac.jp/asia/india/jitnet/ibeo/states.htm

3 Foreign investors have also found that the number of permits and licenses required to operate has been reduced, and 'single-window' investment opportunities have increased.

4 While this may be considered low compared to other emerging economies such as China, it is relatively high compared to past levels of direct foreign investment that India had attracted.

5 Based on information bulletins, press releases and statements of the Embassy of India, Tokyo.

6 Officially defined as expenditure incurred by an individual to obtain a minimum calorie intake of 2,435 in rural areas, and 2,095 calories in urban areas.

7 It has to be noted that poverty per se is not a 'problem' that has to be 'solved.' Poverty in fact is an effect of a number of lopsided programmes and policies that do not tackle the fundamental causes of poverty. These causes are different and require different measures from the level of an individual, household and community to that of a city, state and nation.

8 Source: *India Today*, 24 August 1998, p. 37.

9 A 'positive' outcome of this scenario has been the spurt of petrol and diesel based private generation of electricity, and households and other small-scale users turning to solar and other forms of alternative energy (including bio-gas) for their energy needs.

10 Reforms include irrigation and water supply systems, stabilising markets and food prices, food distribution systems, agro-technology, land ownership patterns and off-farm/off-season economic activities.

11 Statistics for the section on the environment was taken from *The Hindu Newspaper*'s "Survey of the Environment 1998."

12 The World Conservation Union calls for 10 percent of a country to be under natural parks and protected areas.

13 Lecture delivered by Nani Palkhivala on 3 September 1997 in Mumbai at the Birla Matushri Sabhagar.

14 *The Hindu Newspaper*, Friday 15 August 1998, "What should India's Priorities be in the next 50 years?"

15 For example, substantial software investments from multinational and domestic companies can be observed in Bangalore City, the capital of Karnataka State; Hyderabad has similarly benefited in software investments due to the financial climate induced by the state government of Andra Pradesh.

[16] Officially defined as expenditure incurred by an individual to obtain a minimum calorie intake of 2,435 in rural areas, and 2,095 calories in urban areas.

[17] The largely grassroots opposition to the Narmada Hydroelectric project, and other mega infrastructure projects, due to their negative environmental impacts, are examples of such movements.

[18] Highways and other infrastructure projects under 'build-operate-transfer (BOT)' schemes have also been proposed.

[19] *India Today* 7 September 1997.

[20] The provision of Internet services in India, until last year a state monopoly, was recently opened up to private sector participation.

[21] Such services include the development of training curricula and programmes for healthcare, better community health clinics, and better distribution of essential drugs, immunisation, etc.

[22] 'Food insecurity' refers to inadequate and lopsided access to food, low-calorie content, low-income levels and expenditure patterns and related issues.

REFERENCES

Ahluwalia, Isher and I.M.D. Little (1998), *India's Economic Reforms and Development.* Delhi: Oxford University Press.

CMIE (1996) *India Economic & Business Overview.* Bombay: Centre for Monitoring Indian Economy.

CSE (1998), *Towards a Green World: Should Global Environmental Management be Built upon Laws or Human Rights?* New Delhi: Centre for Science and Environment.

Dubey, Amaresh and Shubhashis Gangopadhyay, *Counting on the Poor.* New Delhi: Indian Statistical Institute, 1998

GoI 1998, *India 1997 - A Reference Manual.* New Delhi: Publication Division, Ministry of Information and Broadcasting, Government of India.

Kumar, P. (no date), "Tiger Conservation in India." Document on the Internet. <http://www.cranes.org/PROTIGER.htm>. Accessed 12 January 1999.

Leo's Notes (1997). "Lessons from Economic Growth during 1947-1997." Weekly column in *The Hindu Newspaper.* 15 August 1997.

ORF (1998). *India 1998 - The Observer Statistical Handbook.* New Delhi: The Observer Research Foundation.

Parik, Kirit (1997) *India Development Report 1997.* Indira Gandhi Institute of Development Research. New Delhi: Oxford University Press.

Ramesh, Jairam (1998) "Way to go, India - Economy Globalises, polity regionalises, society localises." *India Today*, 7 September 1997.

SEBI (1997), *Securities and Exchange Board of India - Annual Report 1997* Bombay.

SoE (1998). State of the Environment 1998. Annual Report. *The Hindu Newspaper.*

UNDP (1997). *Human Development Report 1997.* United Nations Development Programme. New York: Oxford University Press.

UNFCCC (1997). *Country Information: India.* Bonn: The United Nations Framework Convention on Climate Change.

USAID (1997). *New Partnerships Initiative: A Strategic Approach to development Partnering.* Washington D.C.: United States Agency for International Development.

World Bank (1998a), *World Development Report 1998.* Washington D.C.: World Bank.

World Bank (1998b). *India: 1998 Macroeconomic Update* (Draft paper) Economic Department Country Studies. Washington D.C.: World Bank Publications.

WRI (1996). *Teachers Guide to World Resources 1994-95, China and India.* Washington D.C.: World Resources Institute.

CHAPTER 10

NEW DEVELOPMENT PATTERNS AND INNOVATION IN CHINA'S ENVIRONMENTAL POLICY

ZHANG KUN AND XIA GUANG
Sino-Japan Friendship Center for Environmental Protection, Beijing

1. NEW DEVELOPMENT PATTERNS AND ENVIRONMENTAL POLICY

There is certainly no doubt that in the twentieth century humanity achieved accomplishments in almost every field of activity. It is therefore completely understandable that people were full of euphoria and pride as the new century dawned. On the other hand, however, in some specific aspects of human development such as the status of our natural resources and the condition of our environment, our situation is much more questionable and our excitement more restrained.

New Development Patterns

The worldwide environmental and ecological crises, which we are facing, challenge us not only with dire warnings but also with unnerving questions. How could our environment, the material base for our human survival, fall into the crisis in which it is today? The answer to this question can no longer be one of responding superficially to each new environmental problem. Our world has traditionally looked upon environmental degradation only as a scientific and technological problem, and the remedy therefore was to find scientific and technological responses. These responses however have been challenged by sad reality, even though we have made great efforts in putting the environment in order.

Our environmental situation has not improved; in fact it has gotten worse, especially in developing countries. A key to this problem is that the creation and

K. Matsushita (ed.), Environment in the 21st Century and New Development Patterns, 239–253.

worsening of environmental problems is very closely related to traditional economic development patterns and strategies. The environmental crisis will not be overcome without a thorough transformation of the traditional development model. In pursuit of this goal, people have long explored various new development approaches based on the idea of long term, environmentally-friendly, natural resource development, which they couched in a variety of catch-phrases such as "organic growth," "all-round development," "simultaneous development," "coordinated growth," and "sustainable development," all of them in fact concrete forms or different stages of what we could call new development patterns.

Environmental Policy

New development patterns have not been brought on uniquely by a renewed awareness of an environmental crisis. Prior to that, many people had become aware that the traditional development model also leads to serious problems in other areas, such as efficiency (competitive ability), equity (contemporary and inter-generational equity), ethics (especially human alienation) (Paoletto 1998). Therefore, the new development patterns strongly imply a basic proposition: that the traditional development patterns must be completely changed, and innovations implemented in every field of human development (Xia 1998).

One significant change in the field of environmental protection during recent years a new tendency to seek a deeper perspective from which to view environmental measures, to look for the causes and countermeasures of economic strategy, in macro policy arrangements and incentive structures. This is completely in line with the spirit of new development patterns, and this change has been reflected as a continuous adjustment and transformation of environmental policies. Environmental policy can be understood as the sum of environmental protection countermeasures within a certain range. The comprehensive reform required by new development patterns, therefore, will manifest itself as innovation and development in the field of environmental protection.[1]

2. FUNDAMENTAL CHARACTERISTICS OF CHINA'S ENVIRONMENTAL POLICIES

Environmental policies are all countermeasures and programs chosen by the State for the purpose of protecting the environment, concrete actions such as laws and regulations, administrative instructions, propaganda, and social supervision, etc. So far, China has developed a fairly complete system of environmental policies, which involve many levels of social structure, and shows diversity of policy instruments.

2.1 Main Components of China's Environmental Policy

China's overall environmental policy covers a broad spectrum of situations including both pollution control and ecological conservation. In this presentation, however, only the former will be introduced.

Pollution control is the main focus of China's environmental policy, a thrust that comprises both control and guiding policies. The control policies are environmental strategies implemented by the State for the whole of society, enforced by legislation, fines, related environmental decrees, etc. The guiding policies or guidelines, on the other hand, include recommendations, awareness campaigns and incentives offered by the State to encourage voluntary public participation in matters of environmental protection, for example 'green' labeling.[2]

2.1.1 Control Policies

The control policies are aimed at the social entities, industries and/or individuals who directly cause environmental pollution or damages. These may include corporations, firms, social organizations, individuals and others. In China, among them, the firms are the main offenders. Most of the control policies have been given legal standing through environmental legislation. The others, such as decrees on meeting regional emission standards or closing certain seriously polluted firms within a limited term, are implemented through direct government action. The following are some specific forms of control policies and environmental management systems (Zhang 1998, Wang 1998).

(a) Environmental Impact Assessment System: According to the Environmental Protection Law and other regulations, construction projects that might impact the environment must go through the formalities of an environmental impact assessment. This would include an assessment on emission levels, protective measures to be taken, a cost-benefit analysis concerning environmental investment, etc.

(b) "Three Simultaneous" Components: In China, the environmental laws are such that the environmental protection components of a construction project must be (1) designed, (2) constructed and (3) in place, simultaneously, at every stage of the project's development. This measure was implemented to ensure that an environmental control investment is made in every case, and to ensure that acceptable emission standards are met after the project starts operation.

(c) Economic Sanctions Strategy: The Pollution Sanctions strategy is both an economic incentive and a system of fines for careless abusers. It has been in operation for a long time in China. It obliges enterprises whose emissions exceed acceptable standards to pay a proportionate fine without excusing them from the duty of controlling their pollution.

(d) Time Limit for Emission Control: Under this system, the government orders those heavily polluting industries or regions to meet emission control standards before a set deadline. The system aims at single polluting sources but may extend to regions as well.
(e) Overall Emission Control strategy: Under this system, the government determines the ceiling of allowable pollutant emissions in a specific region, along with a corresponding plan for emission abatement, in order to avoid the worst possible results in environmental quality.
(f) System of Emission Permits: China's environmental laws require that enterprises or any other bodies which emit pollutants into the environment must report their projected emission levels and apply for an emission permit from the environmental department according to the pre-determined level of acceptable emissions for that locality.
(g) National Environmental Protection Plan and Program: The environmental protection laws provide that any environmental protection plan or long term program must be integrated into the National Plan for Economic and Social Development. Meanwhile the environmental program and total emission control plan in priority areas must be set up and implemented.
(h) Responsibility of Public and Private Officials in Environmental Protection: Officials at all levels are required to sign a responsibility form stating that they will strive towards set goals and targets of environmental protection during their term in office. Their performance results may then be looked upon as indicators of their work achievements.
(i) Close, Stop, Merge or Re-engineer Policies: The government retains the right to close, stop, merge and/or re-engineer enterprises when needed to protect the environment if the pollution situation and the scale of the enterprise warrants it. In certain cases, when the enterprises are causing serious pollution or if they happen to be a small-scale operation within some special industrial sectors (such as in papermaking, etc.), they may be ordered to be terminated.

2.1.2 Guiding Policies

China's environmental policies also encourage voluntary actions of the general public in environmental matters for the whole of society. The environmental policy framework therefore comprises many guidelines, incentives and public education. These guiding policies, in coordination with the control policies, ensure multiple contact points for implementation.

The essential thrust of these guiding policies (listed below) is to initiate and mobilize the ideological or moral resources in the minds of the people to reinforce environmental policies, so that a mutual supporting relationship can be created between the ones that control and the ones that guide. On one hand, the control policies alone would meet tremendous inertia if they were not supported by a sympathetic public, and on the other hand, if the control policies were absent, the guiding ones would be unable to produce essential results.

(a) Assistance to Enterprises for Technological Renovations: The State provides the requirements and directives in terms of environmental protection for companies planning technological renovations of facilities, requiring them to reduce pollutant emissions when the renovation is done. There are usually two ways in which this can be done: by updating technologies and by recycling waste products.

(b) Cleaner Production: Cleaner production has been given much attention in China's environmental policies in recent years. Stimulus may include cleaner production auditing, pilot engineering, cleaner technologies, etc.

(c) Labeling and 1SO14000 Certification: In order to entice enterprises in the production of more environmentally sound products and also to enhance their internal environmental management, the State has begun to promote environmental labeling and ISO14000 certification systems during the mid-1990s. Those enterprises that can meet the two standards may obtain special certifications that could be helpful in their competition with others.

(d) Waste Recycling Incentives: China announced incentives for waste recycling improvements years ago and provided preferential taxation and so on. This policy is now being updated.

(e) Public Participation: The State encourages voluntary social action. The Environmental Protection law reads as follows: "Every organization and individual has the obligation to protect the environment, and also has the right to report and accuse those who pollute and damage the environment". In order to encourage this type of public participation, the Environmental Status Report and the Weekly Air Quality Report have been issued regularly.

2.2 Summing Up Fundamental Characteristics

A summing up of the basic characteristics of China's environmental policy may increase our understanding. The characteristics are mainly as follows:

First, as regards policy orientation, China's environmental policy emphasizes a fair balance between the environment and the economy. China's environmental policy not only takes into account the requirements of environmental protection goals, but also recognizes the burden placed upon the economic system by environmental policies. Generally speaking, how heavy a burden enterprises can bear has always been an important factor when environmental policies were to be set up, thus requiring a mutual compromise between the environment and the economy within the policy framework. Consequently, the top strategy of China's environmental policy stresses coordination between environment and economy, not environmental precedence. An approach based mainly on environmental precedence would probably be unrealistic in a country wishing to promote a very strong economic growth, yet faced with a relatively low per capita level of income. This situation means that China's environmental policy will probably change towards a greater environmental thrust once there is a considerable increase in income per person.

Second, as regards implementation, China's environmental policy displays a fair measure of governmental control. Almost all concrete measures, especially those dealing with environmental management systems are directly handled by the governmental sectors themselves. This kind of administrative behavior throughout the governmental organization lends China's environmental policy a heavy color of government. In recent years, as an example, China conducted a massive campaign for pollution control into the basin of the Huai River and Tai Lake, and the main tools used were the power of the administrative system. In comparison, there are few policies implemented through the social community and by ordinary individuals. For this reason, China's environmental policy is very costly in governmental financial resources, and often creates a sharp conflict with other overburdened departments with limited budgets, as they clamor for financial support.

3. EVALUATION OF CHINA'S ENVIRONMENTAL POLICY

In order to improve and upgrade China's environmental policy, its effectiveness must be evaluated as per results in one main area: the real environmental quality.[3] According to this criterion, the overall assessment of China's environmental policy may be stated as follows: It has achieved much and has basically been successful, but at the same time, certain structural defects and huge difficulties remain.

In terms of effectiveness, the accomplishments that China's environmental policy has achieved during the last twenty years in terms of pollution control and ecological conservation have been remarkable. In the first place, the increasing rate of environmental pollution has somehow been slowed down. From the experience of other countries, wherever there was a very high growth rate in the economy, there was usually a corresponding high deterioration in the quality of the environment. In China's case, the economy and total yearly industrial output increased by an average of 9-10 percent during the last ten years, yet the impact of pollution increased at a much slower rate. It would seem that China basically avoided the usual trend of a corresponding environmental deterioration. Second, China's investment in pollution control increased significantly from 2.5 billion Chinese RMB Yuan in 1981 to 40.8 billion in 1996, while the gross national product increased at only about 0.5-0.8 percent (Institute for Management of China Academy of Environmental Sciences 1998). Third, the pollution levels in several priority areas and some regions made obvious progress. In the Huai River basin, the number of enterprises, which were stopped or closed down reached more than 5,700, along

Table 10-1. Some formal evaluations of China's environmental status over the years

Description	Document	Source
The environment problems we face are still severe, and both the status quo and the prospect of pollution worry us.	Nationwide working review of environmental protection in 1988.	China Environmental Almanac 1990, p. 68. Beijing: China Environmental Sciences Press, 1990.
The environmental situations we are faced with are still severe; in summary, there are improvements in some regions, but the overall situation is worsening, and the prospect is worrisome.	The working report at the Third Nationwide Conference on Environmental Protection, 1989.	China Environmental Almanac 1990, p. 38. Beijing: China Environmental Sciences Press, 1990.
There has been a huge increase of pollutants emitted into the environment, as well as artificially produced damage to the natural ecological environment in certain areas, leading to gradual deterioration of the environment.	The State Council Decision on Further Strengthening the Environmental Protection Work, 1991.	China Environmental Almanac 1991, p. 19. Beijing: China Environmental Sciences Press, 1991.
The large rivers around the country have been polluted to varying extents and the trend is continuously getting worse, especially for aquatic areas near developed industrial cities and towns; noise pollution in urban areas is serious.	Environmental Status Report of 1994.	China Environmental Almanac 1995, pp. 65-73. Beijing: China Environmental Sciences Press, 1995.
Along with economic and population growth as well as the process of accelerating urbanization, the nationwide status of the environment is more and more severe. Pollution in the cities is getting worse and is also spreading into rural areas. The scope of ecological damage is expanding and is grave.	Keynote Report of Fourth National Conference on Environmental Protection	China Environmental Almanac 1996, pp. 16-17. Beijing: China Environmental Sciences Press, 1996.
In comparison to last year, pollution in the cities is worse and is expanding to rural areas; the scope of ecological damage is still expanding too.	Environmental Status Report of 1995	China Environmental Almanac 1996, pp. 88-94. Beijing: China Environmental Sciences Press, 1996.
Due to population growth and economic development that brought huge pressures to bear on resources, the ecological environment in some areas has continually worsened.	Environmental Status Report of 1997	Environmental Networking News, No.7, 1998.
The environmental situation in the whole country is still severe. The levels of environmental pollution and ecological damage in a number of regions have not changed, and some of them are getting even worse; they could now endanger people's health and become factors that hinder economic and social development in certain areas.	Outlines of Countrywide Environmental Protection Work (1998-2002)	Environmental Networking News, No. 10, 1998.

with a corresponding 40 percent decrease in pollution. In addition, the environmental quality in some cities has improved somewhat. Fourth, the number of people involved in environmental protection programs, including management, monitoring, scientific research, education and so on, has increased to nearly ten thousand. (SEPA 1998). Overall, China's environmental policy has shown remarkable effectiveness and the environmental situation would have been much more serious had it not been for this policy.

In terms of difficulties, however, those faced by China's environmental policy should also be mentioned. A key problem is that so far in China, environmental quality has never been given its deserved recognition. Since the 1980s, many official reports about China's environmental situation have been deceptively understated. The most common evaluations expressed guarded pessimism that the environmental situation was "partially improved while overall getting worse" (1988), "the city-centralized environmental pollution problem has deepened, is spreading into rural areas, and the scope of ecological damage is expanding at an alarming rate" (1996), "there is still no change in the condition of environmental pollution and ecological damage in several regions, and some of them are even worse than before" (1998) (Table 10-1). China's environmental efforts are portrayed as continuing along the same line as in the past: that China has a serious problem, and that it is basically being kept under control but not completely. The typical portrayal is that the effectiveness of China's environmental policies have always been offset by newly-introduced environmental pressures, and that her environmental policies have barely maintained the status quo, avoiding any sharp degradation but not having the ability to radically improve the serious state of the environment. This typical evaluation of China's environmental situation raises a fundamental question: Is it time to bring about drastic structural changes to China's environmental policy?

The difficulties facing China's environmental policy seem to center on two areas. First, the decision-making process in economic development matters still does not take sufficient account of the requirements of environmental protection. Although decision-making at the national level has given more consideration to environmental impacts and countermeasures, economic decisions at the regional and enterprise levels have still largely ignored environmental factors. In some cases, this has brought heavy pressure on the environment (e.g., the very serious pollution occurring in the basin of the Huai River, Tai Lake and Chao Lake, etc.). In its goals and principles, China's environmental policy pays serious attention to prevention and the elimination of environmental problems at source, but there is a further need to explore specific ways to implement these principles. Any improvement in the economic decision-making system would be a breakthrough for China's environmental policy.

Second, one realistic difficulty faced by China's environmental policy is the lack of sufficient resources to provide effective supervision and enforce restrictions against environmentally harmful activities. Implementing environmental policies requires a financial commitment, and these costs are often quite high. They may include expenditures in all aspects of environmental management, such as establishing organizations, adding staff members, purchasing facilities, monitoring the results, bargaining with enterprises, legal litigation, etc. The list could be endless. In China, the majority of these costs falls upon the governmental environmental protection sectors at various levels who implement most of these environmental policies. Due to very limited budgets in these sectors, and especially the limited number of people who actually carry out environmental management, these environmental departments are not able to exercise control over the number of enterprises and other bodies who violate environmental laws, with the result that environmental policies are not fully implemented. This is called "slack implementation" in China and becomes more evident as the development of the market economy continues to develop and the number of market entities increases (Xia 1993).

The difficulties encountered by China's environmental policy may also involve other factors such as investment and technology, etc. However, by their nature they are also policy issues. More precisely, they are economic decision-making issues at the macro level at the same time as environmental supervision issues at the micro level. Changes in China's environmental policy may therefore begin with improvements in economic decision-making and environmental supervision.

4. THE DIRECTION OF CHINA'S ENVIRONMENTAL POLICY INNOVATION

4.1 Nature of Environmental Policy Innovation

The transformation of China's environmental policy, as mentioned above, is a kind of structural and fundamental one that can be described as "innovation in environmental policy." Innovation, according to its essential meaning, means a breakthrough from a previous system to gain a new "profit".[4] If we look upon environmental improvement as a kind of profit, environmental policy innovation would, therefore, be a way to achieve it. Environmental policy innovation is a kind of institutional innovation, indicating an improvement of relationship between people in terms of environmental issues. Hence, innovations in environmental policies should follow the same general laws as institutional innovations.

The major general laws of institutional innovation are two-fold: to continually reduce the cost of operation of the institution, and to implement innovative actions with effective incentives (Wang 1992, 1993).[5] Following this principle to assess China's environmental policy, it can be seen that, as a type of institutional system, it has the problem of excessive operating costs added to a lack of sufficient stimulus in favor of social environmental protection. These two issues of economic decision-making and environmental supervision were mentioned above in discussing the difficulties facing China's environmental policy. They can now be analyzed further.

If we take first the lack of coordination between the economic decision-making process and the environmental consequences, we could ask: Why is there such great pressure on the environment from the economic development process in China?

This is because an effective mechanism for coordinating and integrating the decision-making procedures on environmental protection with economic development has been inadequate in China's environmental policy up until now (Xia 1994). This has led to unusually high costs in trying to balance the two, so high in fact as to make any coordination impractical. (Consider for example the situation in the Huai River and Tai Lake basins where economic development efforts are almost completely unconnected to any environmental consideration.) It is obviously difficult for this kind of environmental policy to be efficient.

Now if we look at the issue of insufficient environmental supervision, we find that with the large number of cases that would need to be monitored for pollution abuses, the rather limited financial resources allocated by governments render proper supervision entirely inadequate. Why then are community organizations not getting involved in environmental supervision besides the government? The answer is simple: a lack of incentive.

China's environmental policies, especially the environmental laws, authorize the government to do the job of environmental management, but provide neither incentives nor protection to the general public. Suppose social organizations or individuals would like to challenge a polluting industry. What would they gain from it? (Is the personal interest motive not the basis of a market economy?) And even if they did bring information to the public, they would have no authority to do anything about it. In fact, they might even face retaliation, and be unprotected by law. If nothing is developed in terms of a community's rights and interest to a healthy environment, community and individual resources may never be stimulated or mobilized in favor of environmental supervision, and the "slack implementation" mentioned above may never change.

4.2 Direction of Innovation in China's Environmental Policy

The development of a market economy in China has already promoted an extensive social organization and prompted individuals to get involved in economic activities, but in the field of environmental policy, direct government control remains strong. This definitely presents one clear opportunity for innovation in China's environmental policy today.

The preceding analysis eminently shows that the basic route for innovation in China's environmental policy should be in three areas: (1) improving the economic decision-making process; (2) setting up mechanisms for community awareness and social action; and (3) proper deregulation.

4.2.1 Improving the Economic Decision-making Process

In order to improve the economic decision-making process, appropriate and efficient mechanisms must be set up to help bring about a balanced approach between economic development and the environment. These mechanisms will comprise the following. First, with regards decision-making bodies, special attention should be paid to the concerns of environmental sectors within governments, both at the national and regional levels; and these officials should realize their role, participate and contribute to the environmental decision-making processes. Essentially, integrated decision-making should reflect a kind of relationship between coordinated interests, involving on-going communication and exchange to prevent economic decisions from leaving a trail of serious environmental consequences. Second, all decision-making levels should be involved, especially those at the national and district levels because they may cause even broader environmental impacts than decisions made at the local project level. Third, as for the decision-making format, integrated decision-making mechanisms may take the form of committees, working groups, agreements, votes, round-table discussions, etc. They can all be used depending on the situation. All in all, an integrated decision-making process in matters involving development and the environment is a new kind of approach with long range possibilities and is one of the chief options open to China for environmental policy innovation.

4.2.2 Setting up Mechanisms for Social Mutual Restraint and Balance.

Efforts to set up mechanisms of mutual restraint and balance imply a spirit of public participation and conciliation. However, public action in matters of development and environmental issues should not be confined to mere participation. Rather, they should become a basic driving force of our society. The usual structures that the public can use in expressing their values will probably be social organizations, communities and individuals.[6] The power of social organizations and individuals in environmental policy and implementation have been evident in various countries (both developing and developed countries), wherever they have been provided with supportive policies along with adequate incentives. For example, the damage-related Pollution Compensation System and the Pollution Dispute Resolution System in Japan have proven to be valuable experiences. China's environmental policy urgently needs to extend environmental rights and incentives to the general population so that the public can act as local supervisors and help control environmentally damaging activities. Concretely, their main rights and interests should be as described below:

(a) The right to know: The public has the right to be aware of the facts of a case by access to various information channels and the release of environmental information. Besides the availability of only the Environmental Status Reports and the Weekly Air Quality Reports, environmental information should also contain specifics about certain enterprises (e.g. their production technology, their raw materials, their emissions, etc.). This kind of information could help the public to judge an enterprise's prospects along with what risks are involved with regard to the environment. Then individuals could decide whether to get involved in the enterprise's productive procedures (e.g. to buy its stock) or to campaign against it.

(b) The right to monitor: Environmental laws should empower the public with the right to watch for environmentally damaging activities on an on-going basis. For example, any legal entity and/or individual should have the right to take samples from pollutant sources, request a legal challenge and then, if circumstances warrant it, use these as evidence in a lawsuit.

(c) The right to claim compensation: This right is at the nucleus of environmental rights and interests. Not only those who suffer directly because of pollution but also those who are indirectly affected by the pollution sources should have, according to law, the right to ask for compensation from polluters. This arrangement is the basic motivating force providing the public with a reason to carry out environmental supervision, and is, at the same time, a powerful incentive that polluters can hardly defy. The key here is to make laws, which empower people to take environmental action. (Depending only on an enhanced environmental awareness would stir up only a very limited number of people. Greater involvement can only happen when the incentive of self-interest is present). This point is the most important but also the most difficult one to break through.

(d) The right to participate in some form of economic and environmental decision-making process: The public should be called upon to express their feelings about significant economic policies or the requirements of an environmental impact assessment, especially in cases that affect them, such as construction projects.

4.2.3 Proper Deregulation

In the course of strengthening public involvement, the environmental management instruments usually operated directly by governments would probably benefit if they were simplified. That might include giving up on some environmental systems and administrative measures that are usually inefficient and very costly anyway. Although this option may differ from the view of those who continually seek to broaden governmental functions and powers, it is a view that could have much value. More importantly, this change would not deny the role that government plays in environmental policy; on the contrary, it would maintain a dominant role for the government on a more important side. The real function of government in this scenario is adjustment and control of environmental policies at the macro level (including environmental legislation and economic coordination), as well as impartial verdicts at the micro level. To properly simplify regulations could change the role of government from being a "litigant" in executing environmental policy to that of an "arbitrator," from an "athlete" to a "judge," and this, surely, would be very meaningful (Xia 1997).

5. CONCLUDING REMARKS

These new development patterns present a form of new civilization brought about by a world that faces tremendous challenges based on a new sense of values, whereby society and the economy will indeed be restructured. New development patterns require a thorough rethinking and transformation, and this is sometimes a slow and painful process. This paper discussed one side of the new development patterns, specifically the point of view of environmental policy innovation, in a very introductory manner. The task of perfecting these new development patterns and bringing about environmental policy innovations will be the challenge facing many thinkers during the twenty-first century.

ENDNOTES

1 With regard to "environmental policy," some differences in understanding the term in a broad or and narrow sense have always existed. The broad perspective looks upon environmental policy as all actions and measures taken by the State, including environmental laws and regulations. This paper uses a broad definition of the environmental policy to include in its scope both pollution control policies and ecological conservation policies, and in content formal environmental management systems as well as timely environmental countermeasures.

2 The classification of environmental policy as "control" and "guiding" policies here is for illustration in this paper, and does not imply that such a formal classification exists in China's environmental policy.

3 Xie Zhenhua (Director-general of SEPA, China): "Environmental protection work is just a process. Its performance depends on that whether the environmental quality is improved or not." China Environmental News, 29 October 1998.

4 Schumpeter (1912), the renowned American economist who at an early stage defined and illustrated the concept of innovation as a scientific concept, pointed out that innovation is to introduce a new combination to the existing production system. Here, a productive system can be looked upon as the product of all

"efficiencies"(including environmental efficiency), so environmental policy innovation can be seen as a breakthrough from existing environmental policy. Innovation can be further subdivided into technological innovation and institutional innovation. The former presents a change in the relationship between man and the nature, and the latter between man and man.

5 To continuously reduce the cost of the operation of an institutional system can be understood as to choose a cheaper institutional arrangement with the same effectiveness. Equally, to provide effective incentive for innovative actions also means to let those who create value to obtain the necessary benefits. In China, a planned economy has been replaced by a market economy because the latter is superior to the former in this respect.

6 Social organization mainly refers to non-governmental organizations (NGOs).

REFERENCES

Institute for Management, 1998. *An Economics Study on China's Industrial Pollution Control (Research Report)*, Beijing.

Paoletto, 1998. *New Development Patterns (Draft Report)*. IGES, Japan.

Policy Research Center of SEPA, 1998. *A Brief Analysis for Impact from Various Policies on Cleaner Production (Research Report).*

Qu Geping, 1997. *We Need a Transformation*. Jilin: Jilin Renmin Press.

SEPA, 1998. Trans-centuries Environmental Protection in China, *Environmental Networking News*, No. 8.

Schumpeter, 1912. *Economic Development Theory* (Chinese version). Beijing: Commercial Press. (1985).

Wang Fengchun, 1998. *The Comparison between China's and Japan's Environmental Laws and the Regulations for China's Legislation. (Research Paper).*

Wang Dingding, 1992. *A General Theory on Institutional Innovation*. Beijing: Economics Research, No. 10.

- 1993. Rethinking the Common Procedures of Institutional Innovation. Overseas Chinese Association Economics (edit.), *Efficiency, Equity and Deepening Reform and Open to Outside World*, Beijing: Peking University Press.

Wang Hanchen and others, 1997. *Study on Environmental Policies Framework during the Economic Transformation Term (Research Report)*, Beijing: Policy Research Center of SEPA.

Xia Guang, 1993. *On the Market-oriented Agent System for Environmental Rights and Interests*. China Environmental Sciences, No. 5.

- 1994. Setting up a Normal Mechanism for Integrated Decision-making on Environment and Development. *China Environmental News*, August 1.

- 1997. Flag. Direction. Position. A Related Thinking about the Position and Relevant Policies of Environmental Protection Working. China Environmental News. November 4, 1998. *A Road towards Sustainable Development.* Beijing: Scientific Highlights. No. 3.

Xie Zhenhua (edit), 1997. *Environmental Laws Implementation in China*, Beijing: Hongqi Press.

Zhang Kun, 1998. *The 21st Century Oriented China's Environmental Protection (Research Thesis)* Beijing: Sino-Japan Friendship Center for Environmental Protection.

Section III

New Development Patterns and International Development

Chapter 11

INTERNATIONAL DEVELOPMENT AT THE DAWN OF THE 21ST CENTURY

KEITH BEZANSON
Institute of Development Studies, University of Sussex

1. IGES IN CONTEXT AS A NEW INTERNATIONAL ORGANIZATION

The Institute for Global Environmental Strategies (IGES) is a completely new international organization dedicated to facilitating and supporting human development through environmentally beneficial strategies. An inescapable irony in the timing of the establishment in 1997 is that its birth coincided with a time when global turbulence, instability and discontinuity were probably greater than they have been at any time since the early years of the 1930s. With almost no exceptions, all other international development organizations were established in a much earlier period of relative continuity and stability, in the 'age of internationalism' that had been constructed in the years immediately following 1945. The political and intellectual leaders of the post-war period had been formed by the traumas of economic depressions and World Wars and had united in the common cause of 'never again' allowing them to be repeated.

The international political and economic order that they established depended on United States' economic strength and political leadership, but its foundations went beyond this. They also involved a major change in mindsets - in thinking - and in vocabulary. Central to this was a new idea, the idea that development could and should be universal. Development as a socio-political force was born and "modernization theory" emerged in academe. This idea of universal development was nothing short of revolutionary. The previous mindset had held essentially that most societies were distinct, non-comparable and destined to remain so; to understand most of them was the stuff of cultural anthropology. The new mindset

K. Matsushita (ed.), Environment in the 21st Century and New Development Patterns, 257–270.

placed all societies in the world for the first time on a single continuum - the continuum from least to most developed (defined, of course, by the norms of the most developed). All of humanity became comparable, and all of us were headed to the same place.

The result has been that, for most of the last fifty years, the ideological duel and Cold War were accompanied by an inspiring, publicly-funded experiment in international development, arguably the most noble experiment in human history. Its accomplishments have been remarkable when measured by such indices as life expectancy, infant survival, nutrition levels or even poverty elimination. Its failures have been social disruptions, biodiversity loss and environmental destruction.

Whatever its accomplishments and failings, this great, publicly-funded experiment was built from and depended on a post-war order that has now largely ceased to exist. We have entered a qualitatively new phase in the international system, and the "development" paradigm, in both theory and practice, finds itself largely unprepared for this. It is true that there is the vast, unfinished business of improving lives and meeting basic needs, but current structures, institutions and, most importantly, thinking, are no longer well-suited to meet these challenges. What this calls for, if the still valid objectives and ideals of development are to be achieved, is nothing less than a fundamental re-examination of the concepts of development and progress themselves. If this is not done and done quickly, then the twilight in which the international development effort currently finds itself will be judged by future historians to have been its sunset. It is in this context and against this background that IGES has begun to establish its own agenda.

2. DEVELOPMENT - PAST AND PRESENT

The order that emerged following 1945 defined development in terms of achieving, through international public actions and in the span of one generation, the standards of living that the rich nations of the West achieved in three or four generations, but without the heavy social costs they had to pay or inflicted on others along the way. The ruling dimension became firmly fixed on the view that development was equal to growth in per capita GNP, and the mainstay of early development economics became the Harrod-Domar formula which treated the rate of growth of per capita income as the independent variable. This afforded a universal treatment to all societies which, at one and the same time, divided the world along a sharp North-South axis and established the foundation of thinking about the human condition as linear, predictable and manageable. International development organizations became the embodiment of this universal treatment and, in this context, the new discipline of development economics placed emphasis on 'catching up' through a combination of competent planning, the transfer of Western science and technology, capital accumulation and economic "management" by governments.[1] The simple and easily understandable promise of a better world had a magnetic quality and converted into a towering and inspiring vision which stimulated international enthusiasm.[2]

Enthusiasm for international development as an instrument of public policy has been in decline for several years. In response to this, development organizations - bilateral, multilateral and non-governmental - have been investing in demonstrating and communicating that "development works". Such efforts, their authors hope, will re-galvanize public and political will in industrial countries for increased allocations of public funds for international development. Yet in spite of such efforts, which have now been on-going for several years, the erosion of public support and public finances for international development continues.

This erosion is not to be mistaken as a temporary malaise. Moral indignation and denunciation will not serve to underpin future development cooperation efforts. What is occurring in and to development is but one consequence of a qualitatively changed international order and of the discontinuities that define it. The fact is that the new context for development efforts is greatly changed from that which existed only a few years ago. At a deeper level, the development vision itself, based on Western standards of living to which all humanity was supposed to aspire, has become hazy and blurred. This is part of the larger and vanishing Western ethos of "progress is inevitable" that has been with us since the dawn of the Enlightenment. Western society is abandoning the credo that improvements in the quality of life for future generations are inevitable. The long period of post-Enlightenment faith that the application of reason and advances in science and technology would necessarily lead to improvements in the well-being of people and of the earth itself[3] is receding.

If this assessment is correct, what is called for is nothing less than a fundamental re-appraisal of the very meanings of development and progress at a time of unprecedented turmoil in practically all aspects of human activity. That this will prove no easy task is self-evident, for the fact today is, inside or outside governments, there are no dramatic new ideas that command anything like an intellectual consensus or enthusiastic public support in the fields of money, trade, or development. Yet surely it is the institutions which were set up to bring about development that should be stimulating and leading an effort to achieve the fundamental re-appraisal that is required. For the most part, they are not doing so. It is not so much then that Humpty Dumpty is broken than that he is fast allowing himself to become a museum piece.

The appropriate starting point for any re-appraisal is to place development issues in their new context of which some of the key elements are the following:[4]

(a) The development experiment was anchored in the political order of the Cold War which disappeared as we entered the 1990s. This has been replaced by the uncertainties accompanying a transition to a new, more complex and less predictable world order. Simultaneously, the international economy is experiencing its most profound transformation since the Industrial Revolution, including major shifts in trade patterns, the internationalizing of financial markets, and changes in the nature of work and the impact of technological advances.[5]

(b) At the deeper level of society and culture, the time-honored assumptions that have underpinned the local social order in many parts of the world are being overturned, often with tragic and savage results. This is particularly evident in the developing regions and the former socialist countries, but it is also increasingly apparent throughout Western society. The complex web of human values and interpersonal relations which enable communities to live together is similarly being subjected to unprecedented strains in many parts of the world.[6]

(c) And all of this is driven by scientific advances and technological innovations whose pace and impact - both positive and negative - are unprecedented. As a consequence, those with the capacity to absorb, use, and adapt the advances in science and technology will be better placed not only to enrich themselves but also to influence the conduct and evolution of human affairs. Those unable to harness these advances will be increasingly marginalized.

(d) The assumption of a North-South axis which defines wealth and poverty is firmly imbedded in institutional approaches to development and in the public mind. It fails, however, to reflect a reality where increasing concentrations of individual wealth are found in countries that aggregate statistics continue to treat as poor, and vice versa. The old geographic fault line between rich and poor has been replaced with a near-impenetrable patchwork that cuts profoundly within individual societies and which is making increasingly meaningless the aggregate statistics and concepts which are used to understand development. This, in turn, is changing the spatial nature of society and community. Attitude surveys tell us that perceptions are shifting and that the "included" in Bombay, Sao Paulo, New York and London perceive "their own community" not in terms of the local neighborhood but in transnational terms and also view similarly the "excluded," whether living down the block or half a world apart, as menacing, problematic and even undeserving.

3. RETHINKING DEVELOPMENT AND THE LANGUAGE OF DEVELOPMENT

These, then, are some elements of context. They are but a very partial reflection of a bewildering array of forces, poorly understood and even more poorly managed, that are most frequently described under the single metaphor: "globalization." In his introduction to The Method of Leonardo da Vinci, Paul Valery observed that: "The folly of mistaking a paradox for a discovery, a metaphor for a proof, a torrent of verbiage for a spring of capital truths, and oneself for an oracle, is inborn in us." Yet in all societies, since the emergence of abstract reasoning and language, metaphors, myths and stories have been required to allow humankind to appreciate the complexity of our predicament and to provide guidance for the future. Over the last half of the twentieth century, the idea of a universal and unifying development has been one such metaphor. Its basic premises no longer hold, but the very turbulence and uncertainties that had made this so also make it essential to create new world views and images to guide the efforts aimed at improving the human condition.

As a starting point to accomplish this, the language used to talk about development needs to be modified. This will not be easy, for to quote Keynes: "The difficulty lies not in new ideas, but escaping from old ones". The world has grown accustomed to political and business leaders addressing themselves only to limited manifestations of the problems we confront and the crises in which we find ourselves. This will complicate any rethinking of development. Most official development discourse continues to use the language of unlimited economic growth and expansion, of a North-South fault line separating wealth from poverty, of nation-state roles that no longer exist, and of a "here" versus "there" which no longer accords with reality. This places development in a dangerous situation: unless changed, its language will be judged as incoherent with reality.

The inertia of intellectuals and scholars is perhaps even greater than that of political and business leaders since they are always more inclined to readjust concepts and theories rather than rethink their fundamental relevance and correctness. This is easily understood because, except for unusual circumstances, it is, in general, rational not to cast aside established wisdom, given the huge investments of time and effort involved in mastering such wisdom, and the even larger (and much higher risk) investment required to develop new concepts and theories. A recent example cited with some frequency of such "rational inertia" comes from economics. At the end of the 1960s, it had become clear in the United States that inflation was no longer trading off, as theory held it should, with reductions in unemployment. The initial reaction of most economists was that this "stagflation" was simply not possible, and they worked even harder to demonstrate that the economy would soon return to its "normal" state. Reality, however, proved to be stubborn, and the growing evidence had, finally, to be accepted as a turning point in economics. Ironically, the stagflation phenomenon had been prevalent in developing countries - especially in Latin America - for a full decade before its challenge was obvious in the United States.

The point is that we are dealing with unusual circumstances and that minor adjustments within the frameworks of current constructs of development will likely prove futile, since those constructs take but scant account of the overwhelming new realities which have swept the old order aside. If development thinking is to catch up with those new realities, an essential starting point must be the modification of its language to emphasize, among other things, the diversity and centrality of human values, the expansion of human capabilities, the key role played by the capacity to generate and utilize knowledge and the importance of respecting environmental constraints. Perhaps the very word "development" may have to be replaced by a new one - still to be identified - in order to reflect the variety of new ideas and concerns it should encompass.

Furthermore, the results of any attempt at redefining development and progress can only be tentative and provisional. The need for a (i.e., single) "new development paradigm" has been called for by some, but the emergence of one is most unlikely. There will be no single "Holy Grail", not one comprehensive view that can accord with the human condition in our turbulent times. There is, to use a metaphor, no new Keynes on the horizon. We may yearn for simple and complete mental images (or stories or myths or models or theories) to explain the world around us, but the complexities, paradoxes and uncertainties that accompany globalization, and the profound changes they are exerting on the human condition require that we develop evolving and flexible conceptual frameworks that accommodate blurred boundaries, overlapping categories, ambiguity and uncertainty, while at the same time maintaining intellectual rigor.

4. GLOBALIZATION: ORDER AND DISORDER

It is, then, not just that globalization poses new challenges for international development, but rather that globalization requires new thinking about development itself. It is equally the case that the relevance and value of development thought now depends on whether it influences thinking about globalization. Globalization is the new mantra and, barring an improbable return to protectionism or social collapse, it is not going to go away. It is most often portrayed by its advocates (including many of the major international development institutions) as an integrating and homogenizing force, but the truth is much more complex in terms of its dynamics and impact. First, it is an ambiguous amalgam of potentially contradictory processes which promise both integration and disintegration, inclusion and exclusion, opportunities and problems, equalization and unequalization, creation and destruction, mobility and entrapment, order and instability. It is a deeply uneven process which operates at different levels through parallel processes of regionalization; affects different regions and countries differentially and, within countries, different areas and social groups; and advances at different speeds across economic terrains (for example, more rapidly in financial than in labor flows). It is an unsettling process which imbues economic transactions, particularly in the realm of finance, with greater volatility and threatens to bring more uncertainty and insecurity into people's lives. It is a disempowering process in that it reduces the autonomous power of national governments, undermines the accountability which is the lifeblood of democratic politics and breaks up established social relationships in the realms of kinship and civil society. Significantly, these four characteristics of globalization[7] are evidenced in developing and industrially advanced economies alike.

Both proponents and opponents of globalization appear to agree on one thing: the boundaries and spatial reference points that had long been taken for granted are fast decreasing in significance. What is happening today cannot be understood or explained through the old spatial constructs based principally on the "here" without accounting for the "there" and vice versa. This blurring of the intellectual framework on which the development boundaries of North-South and developed-developing have rested must be expected to increase through, inter alia, the advance of the international marketplace, the continuing spread of new technologies, the functioning of capital markets and the activities of the World Trade Organization (WTO). But it will increase also as a result of the actions, reactions and coping behaviors of individuals, communities and institutions. This duality of forces has been characterized as globalization from above and, simultaneously, globalization from below.

What this suggests is that, far from there being a single "globalization", there are and will be many "globalizations". The phenomenon from which the globalization metaphor derives is based on different and distinct happenings in countless separate places. "Globalization" will almost certainly not lead inexorably to homogenization. Rather, it will continue to produce multiple outcomes, different responses, new institutional arrangements, a complex variety of coping strategies, winners and losers, and new configurations of the included and the excluded. To repeat a point made previously, this is why the redefining of development and progress cannot produce "a new paradigm," and why we must accept overlapping categories, ambiguities, and multiple boundaries.[8]

Many of the bilateral and multilateral institutions of development find themselves today at the forefront of sweeping claims of the benefits of globalization. Other development agencies, mainly non-governmental, have situated themselves in resolute opposition to all aspects of globalization, pointing only to the relationship between globalization and enhanced insecurity, a new and thinly disguised colonialism, financial and social instability, different ways to kill and impoverish, and shifting and usually widening fault lines between rich and poor. In both cases, there is dangerous selectivity and a failure to confront the complex reality of ambiguities, contradictions, winners and losers. The fact is that, for the next many years, the principal agenda for development, both institutional and intellectual, will be firmly established on the axis of this debate about globalization. And this will apply universally - to Britain and Bangladesh, to Berlin and Bangalore. This is more than a rich menu for new scholarship; it is a deep matter of practical challenges for which solutions are urgently required. The credibility and relevance of the international development enterprise will be determined by its response to these challenges.

An essential starting point is for the institutions involved in the study and practice of international development to move quickly away from the current, sterile post-cold war dichotomy of "globalization" versus "localization", and beyond the misleading simplicity of whether the "focus should be local in order to be people-centered" or should "concentrate on trade access through the WTO." What is required is intensive and serious work to increase our understanding of the complex and subtle ways in which the global and the local interact, influence one another and establish both synergies and conflicts in a heterogeneous and fast-changing world.

Local and place-specific assessments of reality are clearly required. But just as "micro"- based knowledge will be imperative, such knowledge will prove of marginal relevance to attempts to improve the overall human condition unless linked with great care to the larger context. The obverse is equally true. Globalization is no more likely to render irrelevant the importance and specificity of place than did the end of the Cold War equate - in spite of eloquent claims at the time[9] - with the end of history. Globalization is far more likely to reinforce the specificity of micro realities, of place, of "here" versus "there". Again, the urgent requirement confronting development theory and practice is to increase our knowledge about and understanding of those interrelationships.

What does all this imply for the institutions engaged in international cooperation? A great deal obviously, but I will limit myself to two major implications.

(a) First, the debate which rests on a presumed dichotomy of micro versus macro or local versus global approaches must be abandoned. It is a false debate which was started and is fostered by development organizations themselves, and it is important that they provide the leadership required to abandon it. These are neither opposing realities nor opposing epistemologies. Attempts to depict them as such fall into that same category of intellectual sterility as the modernity versus tradition dichotomy that characterized early post-war area studies. Serious development thinking and action will need to grapple simultaneously with the interconnection of the general and the specific, to link local experience to global forces, to construct helpful conceptual models on the basis of aggregated local experiences, to test the validity of the premises and assumptions underlying globalization theory and, most importantly, to generate through integrative knowledge systems the new understandings of the human condition that are so urgently required.

(b) Second, development problems today are larger and more complex than those of the past and it follows that the institutions of international development will require a large, varied and complex body of new knowledge if social purposes are to be met. Significant investments and strategic arrangements are required to establish the knowledge and understandings of the multiple interrelationships, differences, transformations of winners and losers, coping strategies and policy alternatives that are required. These understandings involve new issues and relationships of deep nuance and vast complexity, including differing visions of the relationship between nature and humankind, the transformation of markets, new pandemics and the return of ancient scourges, spiritual/religious upheavals and rejections of "modernity", vast flows of international labor and refugees, the unprecedented production of wealth and at the same time increasing social exclusion and poverty. What seems certain is that none of the serious issues that humanity confronts can be looked at uni-dimensionally and, indeed, that the questions of genuine importance have become intersectional.

At a time of considerable turbulence in an earlier part of the twentieth century, the Spanish philosopher, Jose Ortega y Gasset, remarked that: "We don't understand what is happening to us and that is what is happening to us." His assessment is very pertinent to today's forces of globalization which have created a new phase in the international system which is redefining the very meanings of development and progress and which must inevitably result in new concepts and the understanding of a vast new complexity of multiple realities. For most institutions engaged in development cooperation, there is a daunting issue here in that they have been structured to act far more on established policy and practice than to function as "knowledge-based" or "learning" organizations. Yet if they are to continue to function in the qualitatively changed international order, they themselves now require the kinds of radical transformations that many of them have urged for so long on others.

5. THE FUTURE OF DEVELOPMENT COOPERATION

Two questions that needs to be posed directly are, first whether there is a future for development cooperation and, secondly, if there is, is it likely that such a future will be grasped. The preceding argument addresses the first of these questions in the following manner: just as the international order or Pax Britannica collapsed with the first great war of the twentieth century, the Pax Americana that followed World War II has now been replaced by a qualitatively new phase in the international system, the best single metaphor for which is globalization. Development as we have understood and practiced it, and the institutions that were founded to provide development cooperation, were predicated on an international order that has been transformed in major and irreversible ways over the past few years, and the result is that the theory and practice of development now find themselves rooted in an order that no longer exists. This cannot be corrected by actions at the margin, administrative reforms, mergers or increased financing. What is called for is the fundamental rethinking of development and progress, the transforming of development agencies from essentially transactional into knowledge-based institutions and the modification of the very language of development.

The second question is will it happen? There are a number of encouraging signs. The most significant of these is the emergence of new organizations in what is usually referred to as civil society or the third sector. These are springing into being in countless numbers everywhere. Their sizes, forms and purposes encompass vast diversity. Many have moved well beyond the global-local dichotomy and are posing serious questions about how to work at the intersect between the two. They are seeking ways to channel global forces to public good, challenging received wisdom, stimulating fresh thinking on the human condition, and seeking ways to hold to account the powerful transnational energies that have been unleashed. In a prescient 1995 article, Lester Salamon[10] suggested that the current explosion in the third sector will prove to have the same historical import as the rise of the nation-state some 200 years ago. The principal characteristic of the third sector is its vast diversity, but it is exerting a fast-growing influence on the institutions of development cooperation and, more importantly, on thinking about development itself.

In addition, there are some important changes occurring within intellectual circles which go beyond a small number of iconoclasts. As set out in this essay, the intellectual foundations of development remain grounded in the capital accumulation model of economic growth, a genre of economic thinking that has been tightly shut, where GDP is the measure of human well-being and where unfettered and fully open markets (i.e., globalization) are essential to that well-being. For a growing number of economists today, however, the defining traumas are not the wars and depressions of the first half of the twentieth century but the social and environmental crises of the second half which the growth model unwittingly helped to engineer. To these economists, new conceptual and measurement tools are required which treat the market as embedded in the natural and social spheres, and they reject the long-standing approaches to the latter as "externalities."

This questioning of the growth model has been most closely associated with environmental approaches and such writers as Anil Agarwal, Kenneth Boulding, Salah El Serafy, Herman Daly, Hazel Henderson, Robert Repetto.[11] More recently, however, a parliamentary report in France called for new indicators of progress. A similar call has been made by the Treasury of Australia. And a new organization in the United States, 'Redefining Progress,' comprised mainly of economists, was founded in 1994. Its declared purpose is to act as a catalyst to a thorough rethinking of economic policy and its underlying premises and to stimulate broad public debate over the nature of economic progress and the means of attaining it.[12]

There are also stirrings of change from within the transnational financial and business community itself. Some of the central bankers in G7 countries have voiced concerns publicly that the financial economy has become dangerously detached from the real economy and the well-being of people. The magnitude of costs associated with an avalanche of natural disasters, the cause of which for many is climate change has transformed many large insurance companies into environmental advocates. The World Economic Forum (Davos) meetings of the past two years have included significant expressions of concern over the growing disparities of income that are accompanying globalization. The signs here are, however, far from mainstream. Questions and doubts by international business leaders about certain aspects of globalization are nevertheless surfacing and there have even been some appeals for a fundamental re-appraisal of the meaning of development. The most famous recent example of the latter appeared in an Atlantic Monthly article written by George Soros[13] in which he recalls Karl Popper's The Open Society and Its Enemies, which drew attention to fascism and nazism as antithetical to human development. Soros goes on to argue that today's enemy is "laissez-faire capitalism" and that we need urgently to re-think both globalization, to establish new arrangements to channel its raw energies, and to re-define what we mean by progress. The point here is not to present the Soros argument as valid (it has been attacked strongly by many), but rather to note that the author of this viewpoint is George Soros and that his article appeared in the Atlantic Monthly.

These then are a few indications that a rethinking of what is meant by development is occurring. Confronted by the fractured global order brought on my globalization, however, the array of institutional inertias that have been built into the international system over the last half of the twentieth century are focusing principally on institutional self preservation. The weight of current evidence suggests that the prevailing mindset of international cooperation institutions remains fixed more on advocacy, moral indignation and denunciation than on coming to terms with the contradictions now inherent in the practice of development cooperation and in its intellectual underpinnings.

This may be explained partly by the ferocity of some responses to suggestions from international development institutions that the environmental or social costs of globalization are unacceptably high or that new governance mechanisms are needed to direct its energies. There is perhaps no better example of this than the response of influential US congressmen to a serious and scholarly examination by a number of highly respected economists under the United Nations Development Program (UNDP) auspices of possible taxation of international capital flows, particularly the Tobin tax.[14] Regardless of the intellectual merits or failings of that examination, the very idea that an "official" organization of international development would associate with thinking about taxation of the unrestricted flow of global capital was anathema to these powerful politicians. They made clear that, should the scholarly discourse be published by UNDP, funding in support of the work of the organization would be affected.

There is also the factor of time. It has become conventional to refer to the much quoted formulation of a "paradigm shift," set out in Thomas Kuhn's The Structure of Scientific Revolution. But what is usually overlooked by those who invoke the "paradigm shift" is the very long time period which Kuhn concluded was required for genuinely new thinking to gain currency. To make this point, he quoted the physicist Max Planck: "A new scientific truth does not triumph by convincing its opponents and making them see the light, but rather because its opponents eventually die." This suggests that the institutional rethinking of development and progress will require a very long time indeed.

On balance, the above factors suggest that the development experiment as we knew it over the last half of the twentieth century may well be in its twilight. A great many of its moorings have disappeared with the magnitude, scope and intensity of changes brought on by globalization, and, if it is to catch up with or lead the change that is required, it will need to remove its own numerous and extensive inertias. A former high level official of the United Nations made the following comment about the experiment in development cooperation:

An important experience, without precedent in modern history, is coming to an end. It will have lasted much less than was expected. Born in the midst of contradictions, it dissipates in ambiguity. It means renouncing an ambitious but ill-conceived enterprise. Its original noble intentions have been progressively submerged by other considerations which, inevitably, have led to mutual recrimination and disillusion.[15]

Tibor Mende made this statement in 1972 when reflecting on the failure of development assistance to live up to the high expectations it had raised. He was also concerned that it was an instrument of the Cold War and that it might become a new form of colonialism. If his announcement of the demise of the development experiment proved premature in the early 1970s, might this not also be the case at the end of the twentieth century? The obvious fact is that Mende made his prediction when the post-war order was still very much intact and the experiment in development cooperation was part of, and suited to, that order. The developmental contradictions and ambiguities to which Mende was referring were but sub-components of the contradictions and ambiguities of the larger international order. That is no longer the case and this greatly increases the probability of accuracy for any similar prediction made today.

Yet motivations to deal with poverty and misery will surely persist, as will the desire for an overarching vision of a better, fairer and safer world. I am convinced that over the twenty-first century the centre of the world's political stage will be taken up with the relationship between rich and poor, with the problem of income distribution within nations and communities, and, to an extent unknown in the twentieth century, with issues involving the diversity and centrality of human values and non-material aspects of well-being, including our physical environment. The task is to intensify the quest for paths by which economic inequalities between and within nations might be reduced, to spread the gains of globalization while preserving and healing the environment and to reaffirm the centrality of human ideas, ideals and foresight.

Taking the above factors as a whole, however, I am personally sceptical that the organizations of international development will rise to such massive challenges. What seems more likely to me is that we are witnessing the requiem of development experience we have known and practiced over the last few decades. In this regard, the probabilities are high, in my view, that most international development institutions will go out with a whimper and with a good deal of whining. That, however, is not the end of it, for the history of human civilization, at least from the Hellenic period, confirms that there is a magnificent indestructibility about human hopes and aspirations and especially aspirations for liberty, social justice, material progress and spiritual fulfillment. Such hopes are themselves a part of our future and will help to determine its features. Because of this and in place of the great development experiment of the past fifty years and of most of its current institutions, this century will see a redefined development at the centre of human endeavor, together with very different structures and institutional arrangements that will correspond with the qualitative new phase in the international system and with the spirit of new times. IGES as a new institution can play a central role in bringing this about.

ENDNOTES

1 This "paradigm of development" was questioned by some, including Dudley Seers who in 1969 attacked the equating of development with economic growth and argued strongly that it was misleading to think about and practice development on the basis of a North-South divide.

2 This enthusiasm, however, did not result in the meeting of financial targets such as 0.7 percent of GDP for development cooperation and as early as 1969 the Pearson Commission noted and lamented the arrival of "aid fatigue" (*Partners in Development*, Report of the Commission on International Development).

3 Both public opinion surveys and a wide array of sociological studies confirm that, throughout Western society as a whole, it is no longer believed that succeeding generations will experience improvements in quality of life. Indeed, most results indicate that it is widely expected that the next generation will experience sharply reduced standards of living.

4 The list that follows is extracted from an essay by Keith Bezanson and Francisco Sagasti, *'The Elusive Search: Development and Progress in the Transition to a New Century,'* Ottawa, International Development Research Centre, May, 1995.

5 In an issue which focused on the meaning of these changes, *The Economist* magazine concluded that, early in the next century, nine of the 15 leading economies in the world will be those countries we now call "developing", that China will replace the United States as the world's largest economy, that India will replace Germany as the fourth largest, that 85 percent of the world's population will live in the South, and that, before the end of the current decade, 17 of the world's largest cities will be in "developing" countries.

6 Samuel P. Huntington's analysis of the unfolding world holds as its central proposition that modernization and globalization are causing people with similar cultures to come together and those with different cultures to come apart. See Huntington's *The Clash of Civilizations and the Remaking of World Order*.

7 I am grateful to my colleague Gordon White of the Institute of Development Studies for having drawn out these four characteristics.

[8] For a helpful synopsis of such issues, see Ronald Dore, 'Convergence in Whose Interest?' in *National Diversity and Global Capitalism*, pp. 366-374.

[9] See, for example, Frances Fukuyama's *"The End of History and the Last Man"*.

[10] See Lester M. Salamon, *The Rise of the Nonprofit Sector*.

[11] For a useful overview of such thinking, see Hazel Henderson's *Paradigms in Progress*.

[12] See Cobb, C. Halstead, T., and Rowe, J., *If The GDP Is Up Why Is America Down*.

[13] See Soros, G., *The Capitalist Threat*, Atlantic Monthly, February 1997.

[14] Based on a private communication from Mahbub ul Haq who organized and co-chaired the examination.

[15] Quoted in Helan Jaworski in his *Hacia Nuevas Formas de Relacion con el Sur del Mundo*, Vitoria-Gasteiz, Servicio Central de Publicaciones del Gobierno Vasco, 1993, pp. 27-28.

Chapter 12

ENJOY IT BY GIVING IT UP - Toward Sustainable Development Patterns

KIRIT S. PARIKH
Indira Gandhi Institute of Development Research

Present development patterns raise questions about their sustainability. We see environmental degradation and ecological destruction all around us. The global commons are being overloaded. The ozone layer hole, the climate change threat, the pollution of oceans and the extinction of many species are facts we cannot ignore. With increasing population and growing prosperity, the present development patterns will further stress our environment. These patterns seem to be environmentally unsustainable.

Besides being unsustainable from an ecological perspective, the present patterns and their likely consequences are perhaps socially and politically unsustainable as well. Our present development patterns have to change. What new patterns are desirable? How can we attain them? What questions would such a transition raise? Do we know the answers? In what way can research help in answering them? These are the questions addressed in this paper.

In order to emphasize that we are concerned with development patterns and not with development itself, our first section will look at the need for development. We will then examine the present development pattern and its characteristics. These will be followed by a look at the forces that make the current patterns unsustainable. In Section 4, we explore characteristics of some desirable development patterns. Finally we conclude about the implications for policy and research.

K. Matsushita (ed.), Environment in the 21st Century and New Development Patterns, 271–290.

1. NEED FOR DEVELOPMENT

A large part of the world needs development in whichever way one defines it. There is still much chronic hunger, poverty, illiteracy, ill-health and human misery in the world. In 1993, some 1300 million persons, nearly 30 percent of the population of developing regions, lived on less than $1 a day. If poverty continues to grow at the current rate, by 2015 there will be 1.9 billion persons living in poverty. Clearly, a more rapid rate of economic growth is needed in the developing countries.

Other indicators of development also lead to similar conclusions. In low income developing countries with a total population of some 3 billion persons, adult (above 15 years) illiteracy is 45 percent, but if one excludes China with its adult illiteracy rate of 27, the rate exceeds 60 percent. Even in terms of primary school enrolment, in many developing countries, the ratio is only 60 to 70 percent. Thus adult illiteracy will persist for years to come. These countries need to devote much greater resources of money, people and management to education. This will also result in accelerated economic development.

In terms of health, infant mortality in developing countries was 59 per 1000 births in 1993 while the mortality rate of children less than five years of age was 80 per 1000 children. Life expectancy at birth in 1996 in low income countries was 64 years (58 excluding China and India) compared to 81 for high income countries. In order to provide better health care, as well as to increase access to sanitation and a clean water supply, also developing countries need additional resources that economic growth could provide.

We look forward to a society where every human being is free from hunger and want, has freedom to decide what he/she wants, has the knowledge and good health to pursue personal goals and realize them, to develop one's full potential, giving full scope to his or her individual creativity, and live a life of happiness and fulfillment. Even if we look at the wider concept of development, we can see that more development is needed rather than less. The present pattern has led to some unwanted problems. We need to discover how our ways can be modified to minimize the unwanted consequences, but we must not throw away development itself to get rid of some undesirable side effects.

2. PRESENT DEVELOPMENT PATTERNS

Before we can think about new development patterns we need to understand what is the current pattern. What are its characteristics? Which of these need to change?

2.1 Mass-Markets and Technology Drivers

The main assumption behind the present development pattern is that it is possible to maintain an economy that provides a high standard of living for every individual in the society. This high standard of living is reflected in the consumption of goods and services along with the opportunity people have to choose what they want. Such an economy is sustained by mass consumption and is geared to meet mass consumption. To sustain growth in such an economy, one needs a growing demand which will be ensured by increasing the variety and improving the quality of goods and services. This gives an advantage to large firms who can exploit the economies of scale in production and innovation. Thus two important characteristics of a modern economy are that it is mass market driven and it is technology dominated.

How long can demand keep growing? Since the bulk of the global population has a level of consumption that is way below that in industrialized nations, the scope for growth is very large. But even in industrialized countries, demand keeps growing.

2.2 Dominance by Few Large Firms

Human want is insatiable. Man's greed knows no bound. As an old Sanskrit proverb says:

"a pauper pines for a hundred
the centurion wants a thousand
one with a thousand aspires a million
and a millionaire wants nothing less than
to be the lord of the universe."

(This saying is a more than a thousand years old, so an adjustment is needed for inflation.)

This potential for endless demand is realized through advertisement and other forms of persuasion. A consequence of ever-increasing demand is that manufacturers build obsolescence into their products. This increases the material intensity of the economy and also adds to solid waste generation.

The insatiability of wants is not just confined to material goods. It is also for power and dominance. Firms want to be ever bigger not just because economies of scale dominate decisions but also because their chief executives want to be number one.

Thus another important characteristic of present development patterns is dominance by a few large firms with the power of monopoly.

2.3 Countering the Power of Private Monopolies

Societies have recognized that the power of monopolist capital needs to be contained. One response was to have public monopolies instead of private monopolies. The extreme form of this is communism where the state owns all capital. Other responses have moved towards regulation by the government of restrictive and monopolistic behavior.

Public ownership of capital has not worked well. It has led to inefficient, material intensive and environmentally insensitive production systems in many communist countries. The performance of publicly held monopolies in non-communist countries have also been, by and large, unsatisfactory. Thus privatization is now in vogue.

2.4 Labour Unions - A Mixed Blessing

The pursuit of private profit by capitalists, even where they are not monopolists, often results in pressure to reduce labour costs, and in so doing they tend to exploit labour (i.e., pay less than its marginal product). Labour has countered by getting unionized. This has prevented much exploitation of labour.

However, big private capital and big labour unions learned to have a cozy relationship at the expense of consumers. In many countries, this has led to a consumer movement whose objective is to protect the rights of consumers. This has provided an effective third force in many situations. However, in some countries public sector labour unions wield enormous power and almost hold the society to ransom. Thus for example, in India public sector employees constitute a labour aristocracy. They work less, have more benefits, have lifetime job security and get paid three times as much as workers in the private sector. Naturally many of these public sector enterprises register losses, but the political power of the unions is such that privatization or reforms are stalled. The rest of society pays the price. These kinds of mismanagement hurt the poor in two ways. On the one hand, government finances the losses of the public sector by cutting social expenditures for health, education and anti-poverty programmes. It also reduces growth which reduces opportunities for employment. On the other hand, it also slows down the growth of employment inasmuch as the fear of adverse legislation, enacted under the political pressure of public sector unions unduly protecting labour, encourages entrepreneurs to opt for more capital intensive technologies.

2.5 Growing Urbanization

Growing industrialization, which increases demand for the variety of products has resulted in growing urbanization. A large city provides a demand for the output of industry. It attracts a large pool of skilled labour, a variety of suppliers for the many parts, components and inputs that may be needed, well-developed financial services as well as the social infrastructure of good schools, good medical care, and variety in entertainment. The economic advantage of locating in big cities seems to be large enough to compensate for the inconveniences of commuting and congestion. Urbanization is thus a dominant characteristic of today's development pattern in almost all countries.

2.6 Infrastructure and Social Capital

Industrial development is facilitated by - and in fact, pre-requires - the development of a complex infrastructure: transportation, communication, power, law and order, a legal system for the settlement of disputes, as also education and health care. In many countries, these are provided by the state. They either provide large economies of scale and involve heavy investments with a long gestation period, or they have externalities that a private investor cannot easily capture. At the same time, given the externalities, they are socially very desirable and productive. Thus present development patterns anticipate an active role of government in providing infrastructure and social services.

2.7 Welfare State and Social Security

The increasing productivity of industrial economies has made it possible for societies to institute social security measures. Such security is necessary for the welfare of people inasmuch as they are vulnerable to uncertain events outside one's sphere of influence. The fear of such misfortunes can cause much suffering. Protecting people against these risks also facilitates job mobility and increases flexibility of the industrial system, thus leading to greater efficiency. Governments now consider it their responsibility to look after the welfare of their citizens and provide social security. In that way, most states have now become welfare states to some extent or other. Many provide unemployment insurance, health care and support for education. Yet there is a down-side to this: generous unemployment benefits can reduce the incentive to work. While a government's responsibility to ensure social security is not denied, the extent of support and method of providing it are now discussed. More efficient ways that reduce misuse and reverse incentives are now being explored.

2.8 Summary of Present Patterns

Present development patterns based largely on industrial development are:

-mass-market driven,
-technology dominated,
-resulting in growing urbanization, and
-dominated by few large firms.

To counter the monopoly power of firms, societies have:
- -instituted regulatory mechanisms, and
- -public sector ownership of key sectors;
- -organized labour unions, and
- -formed consumer interest groups.

At the same time, governments facilitate development by:
- -investing in infrastructure,
- -promoting education,
- -providing health care, and
- -instituting social security systems.

Development in some countries is being held back by:
- -inefficient public enterprises operating at a loss; and
- -labour unions resisting reform.

3. CHANGING PATTERNS

Developmental patterns are changing under a number of influences. First, our improved understanding of how development and economic growth take place suggests new policies and processes which in turn affect the pattern of development. Second, technological progress is rapid and accelerating. The information revolution has already altered the way societies function. It will continue to do so. It also opens up new possibilities and opportunities. Third, the world economy is becoming more integrated. Globalization has profound impacts on patterns of development. Fourth, the environmental consequences of present patterns of development strongly suggest that these patterns are unsustainable. They must change. We now look at each of these influences in turn.

3.1 Changing Paradigm of Development

Development is more than economic growth. It means increasing people's choice to live the kind of life they want and their ability to do so. While economic growth is important, it is an instrument and not the final goal. So, while education improves human skills and leads to growth of human capital, education's major importance is that it improves peoples' capacities to do what they want. This is more than a semantic difference and it could have impact upon patterns of development.

Putting people first also means much greater participation of people in decision making. The scope for an all-powerful, all-knowing central planner is no longer there. It also means change in priorities and hence in patterns of development.

The new paradigm is not merely a philosophical change, it has evolved from the failures of past efforts of development. These failures have also taught us some lessons about development.

(a) Top-down approaches to development have had limited success in achieving holistic development. The failure is at various levels and for similar reasons.

(i) The most notable shortcoming is in poverty reduction programmes. Poverty has persisted in many countries despite various governments' stated objective to reduce poverty by their development efforts and despite their allocation of resources for it. Often these efforts have not addressed the most relevant problems of the poor who have generally not been consulted in defining the problem. Thus "their reality" has not been sufficiently reflected in the programmes. Participation of the stakeholders, the beneficiaries, the poor, is now considered essential in order to pose the relevant questions.

(ii) Participation of the stakeholders in the design of the programme is also important to find feasible, implementable solutions. Their knowledge of the local environment, their appreciation of implicit constraints and awareness of opportunities can make critical contributions.

(iii) Stakeholders' co-operation is also essential for successful implementation of programmes. This is particularly so for projects relating to natural and environmental resources. Such co-operation is easier to obtain if the stakeholders have participated in problem identification and initial design of the project so that a sense of ownership can be created.

(iv) Stakeholders' participation can also be facilitated by the involvement of civil society institutions and non-governmental organizations (NGOs).

(b) Even when they are well-designed and have addressed relevant problems, top-down centralized approaches have degenerated due to capture by bureaucrats and/or vested interests other than their intended beneficiaries, as well as from problems of leakage and corruption. To guard against corruption and inefficiency, we need transparency and the involvement of civil society institutions, which perform the functions of a watchdog.

(c) While government has to play an active role in development, its appropriate role is now considered very different than the role governments have played in many countries. The inefficiencies introduced in the economy due to government failure (i.e., the inability of the policy making and administrative apparatus of the state to perform as per the stated social objectives) are often much more harmful than the inefficiencies brought about by market failures that the government policies were intended to correct. Governments' role is now seen to be at once limited and enlarged. In economic matters it is now widely believed that governments should confine themselves to using market friendly instruments to ensure that markets function competitively and to regulate monopolistic behavior. In the area of education and other social infrastructures, governments should be more active than some of them have been in the past.

(d) The private sector has to play an important role in economic development. The economic efficiency with which private sector operates could be mobilized to accelerate economic growth. However, it needs to be regulated to ensure that social objectives are not compromised. Such regulation should as far as possible, rely on market friendly instruments.

Apart from the lessons of experience during the last five decades, one needs to recognize the changing environment within which countries must operate today. Among these are the accelerating pace of technological change, increasing globalization and the mounting environmental problems in many countries and at the global level.

3.2 Accelerating Technological Progress

Technical knowledge has become an important factor of production. A firm's competitiveness often depends on its command over it. A country's competitive advantage also rests on its ability to create an environment in which creativity can flower and innovations flourish.

It also requires an entrepreneurial system which takes the innovations and uses them in economically profitable ways.

Knowledge accumulates. Innovations depend on the stock of knowledge. The stock of knowledge is growing rapidly and it is accessible to ever more people. The number of people trained in science and technology, or those doing research is larger today than all researchers in science and technology that ever lived on earth. Consequently new ideas, new products, new processes, new innovations and new inventions are coming out in growing numbers.

As a consequence of the increasing pace of technical innovations, products become obsolete long before they are worn out. The product cycle becomes shorter and shorter. Firms have to recover their investment in research and retooling in a shorter and shorter period. This in turn requires a larger market. Globalization becomes a necessity.

The information technology that has swept the world in the last decade is also changing the world in profound ways. It affects production processes. Automation is now ubiquitous. Computer chips are everywhere. It has made information a most important factor in economic activities. Information provides a competitive edge to firms. A firm which knows more about consumers is able to market more effectively. With better information about suppliers and their availabilities, a firm can cut the cost of inputs. With greater knowledge about future possibilities a firm can take a more effective hedging action. Investors in stock markets who gather more information about firms can make a better investment decision and earn a higher rate of return. However, the cost of collecting information does not depend on how much one has to invest. This cost remains practically the same for one who wants to invest 10,000 dollars or for one who is prepared to spend 1 million dollars. Naturally, the one who has a million dollars to invest would collect more information and will get a higher return. Thus information provides economies of scale to firms in production and marketing as well as to investors. This then leads to larger firms, global firms. We have seen how, even in software production for example, which is almost like a handicraft, Microsoft has come to dominate the scene.

3.3 Globalization - Inevitable and Inescapable

The communication technology which beams the same programmes all over the world can reach a large part of the world's population. The World Cup soccer finals in 1998 were watched by billions of people around the world. This integration brings awareness about what is possible and what is available, and people learn to aspire for new products and pleasures. A global market has developed.

Table 12-1. Increasing trade in a globalizing world.

Region / Country	Trade % of PPP GDP 1986	Trade % of PPP GDP 1996	Trade in Goods % of PPP GDP 1986	Trade in Goods % of PPP GDP 1996
Low Income	7.1	7.9	33.8	56.9
Excl. China and India	12.0	15.7	50.8	92.4
China	6.6	7.1	35.5	58.4
India	3.9	4.5	16.4	31.1
Lower Middle Income	12.5	20.0	47.1	84.5
Upper Middle Income	12.5	24.1	59.0	77.6
High Income	26.5	38.9	70.4	178.8
World	20.7	29.1	63.8	93.8

Sources: Tables 6.1, World Bank (1998), World Development Indicators (1998).

Globalization is now an inescapable reality for any country. It is a given that the communication revolution has turned the world into a global village. Just as everyone knows what is going on and what is available in a village, so is the case in this global village.

Thus the importance of trade is increasing for most countries and will continue to do so. The economies of the countries of the world are getting integrated into a global economy.

This is clearly seen in Table 12-1. Trade in goods as a percentage of GDP generated in goods production has increased from 64 percent to 94 percent over the decade 1986-96.

For high income countries, as Table 12-1 indicates, this ratio has grown from 70 to 179 percent, which illustrates how globally integrated these economies have become. The emergence of a global market and multi-national firms have also led to a deeper integration of production. Parts and components are sourced from many different countries. Intra-firm trade has become a large part of the global trade. A modern U.S. car or a computer may have components coming from scores of countries. The ease of communication and ubiquitous large computing power have made possible such a deep integration. In fact, this type of integration is even

Table 12-2. Growing importance of private capital.

	Net Private Capital Flows			Foreign Direct Investment		
	$ Billions		% of GDP	$ Billions		% of GDP
Country Groups	1990	1996	1996	1990	1996	1996
Low Income	11.6	65.2	4.1	4.7	49.5	3.1
Middle Income	29.2	181.8	4.4	19.0	69.4	1.7
High Income	-	-	-	167.9	95.7	0.8

Sources: World Bank (1998), World Development Indicators (1998).

necessary to preserve a competitive edge. Computers have also made it possible to manage larger and larger firms with efficiency. The diseconomies of scale have been pushed back.

Globalization has become most visible in the financial markets. The flow of financial capital across countries today is mind-boggling. Private capital flows for direct investment and portfolio investment flowing into developing countries have grown rapidly from $25 billion in 1990 to $150 billion in 1997. The importance of these flows can be seen in Table 12-2 which shows private capital and foreign direct investment (FDI) as a percentage of GDP for different groups of countries. For low income countries, net private inflows have reached 4 percent of GDP. With a reasonable capital output ratio, this should mean an additional growth rate of 1 percentage point.

Foreign exchange trading has seen a dramatic growth from $10 billion to $20 billion a day in 1973, to an average of $1260 billion a day in 1995; compared to world trade, this is 70 times as high.

These short term capital flows provide firms with ways to hedge against uncertainties. Yet if firms are not careful they can get into trouble. The Barings Bank episode, in which unmonitored trading by one employee resulted in the collapse of this historic bank in 1995, is an example. Even a country can get into trouble as we have seen in financial crises which unfolded in East Asia during the late 1990s. The same flows that helped countries develop rapidly can be the cause of a meltdown if domestic financial institutions are not properly supervised,or if short term borrowing is coupled with a term mismatch between borrowing and lending. Integration with a global economy provides the country opportunities for efficiency and growth. However, it entails its own risks. The country gets exposed to international shocks and even policy mistakes of other countries. Does global integration improve human welfare? Do the gains of growth more than offset the pains of added volatility? Are there ways in which a country can globalize while shielding itself against the ill-effects? We still don't have clear, generally accepted answers to these questions, and globalization is in transition.

Apart from the risks of greater volatility, globalization is leading to some other problems too. At present only a handful of developing countries benefit from FDI. The top 10 developing countries garner some 73 percent of FDI while the remaining developing countries, comprising 45 percent of the population of low and middle income countries, get only 27 percent.

In 1997, Sub-Saharan Africa got only $3 billion of the $121 billion of FDI going to developing countries. Many countries are thus completely excluded from this benefit of globalization. These countries continue to rely on aid which unfortunately is dwindling. official development assistance (ODA) including non-concessional finance has come down from $63 billion a year in 1990-91 to $40 billion in 1996-97. Unfortunately governments of rich countries are abdicating their aid responsibility to private capital, which mostly goes to where profits are high. That excludes most of the least developed countries.

The process of globalization has the potential to change the world. It can lead to a widely shared inclusive prosperity or a highly uneven development from which a significant number of countries and people are excluded.

3.4 Environmental Consequences: Growing Constraints

The present patterns of development have led to such environmental problems that these patterns have to change. The environmental consequences of development are visible everywhere

Table 12-3. Extent of dependence of poor and wealthy households on common property resources (CPRs) in dryland India.

STATE	HOUSEHOLD CATEGORY	CPR CONTRIBUTION TO INCOME (percent)	FUEL SUPPLIES (percent)	ANIMAL GRAZING (percent)	DAYS OF EMPLOYMENT PER HOUSEHOLD
Andhra Pradesh	Poor	17	84	-	139
	Wealthy	1	13	-	35
Gujarat	Poor	18	66	82	196
	Wealthy	1	8	14	80
Karnataka	Poor	20	-	83	185
	Wealthy	10	-	29	34
Madhya Pradesh	Poor	22	74	79	183
	Wealthy	2	32	34	52
Maharashtra	Poor	14	75	69	128
	Wealthy	1	12	27	43
Rajasthan	Poor	23	71	84	165
	Wealthy	2	23	38	61
Tamil Nadu	Poor	22	-	-	137
	Wealthy	2	-	-	31

Source: Jodha N.S. (1991).

In developing countries, problems of local pollution are widespread. Air and water pollution, soil degradation, degradation of community property resources (CPRs), such as woods, forests, lakes, fisheries, etc., adversely affect the welfare of the people and particularly the poor. This environmental degradation has high, and in many countries, rapidly growing costs. It considerably reduces the gains from economic development.

When CPRs are degraded, the poor get hurt. How important are CPRs for the poor? Systematic data from large sample surveys are not available to say how much of their consumption is obtained from the CPRs by the poor. However, there is some evidence from small surveys. Thus Jodha (1991) found from a survey of 82 villages in India that the poor obtained 66 to 84 percent of fodder from CPRs in some states. Moreover the CPRs provided 137 to 196 days of employment per household and 14 to 23 percent of the income of the poor. (See Table 12-3.) Jodha also found that the average area of common land declined by 31 to 55 percent between 1950 and 1985.

A more recent systematic survey in Bangladesh carried out in 1989-90 (Hossain, 1995) shows that the rural poor in Bangladesh obtain as much as 22 to 28 percent of their market consumption from community ecological resources. This is not counting the value of fish obtained from common access waters which is significant for Bangladesh.

A good indicator of household dependency upon CPRs is provided by the use of fuelwood. It is estimated that 298 tonnes of fuelwood were used in India in 1991, much of this by individual households. Fuelwood provided 50 to 60 percent of the domestic energy consumption of rural households. The officially recorded felling of wood is only 20 tonnes , although official data do not include illegal felling by large-scale contractors. However, most of what they log is used as timber not as fuelwood. Thus at least 150 million tonnes of fuelwood are collected in India every year and burned. Where is this fuelwood coming from? Despite uncertainties of the source of fuelwood, the importance of CPRs for the poor is obvious.

In Africa, a similar situation exists. Fuelwood provides 35 percent of energy for rural households. The total fuelwood consumption is estimated to be 127 tonnes in 1991, most of it collected by households from CPRs and forests. Even in towns, the poor rely on fuelwood. In five towns in Kenya more than 80 percent of the households constituting the poorest quintile use fuelwood for cooking, and on average, 35 percent of all households use fuelwood for cooking (Soussan, 1987). More than 90 percent of the poorest quintile households cook with biofuels, and even among the richest quintile households, 60 percent use biomass fuels for cooking. The corresponding figures in Vietnam are more than 50 percent and 60 percent respectively. (Reddy, Williams and Johansson, 1997).

When the poor rely on natural resources for more than 20 percent of their consumption, any degradation of natural resources means lower consumption by the poor. This aggravates poverty and increases the poorer population's compulsion to find other sources of natural resources, even if it means going farther afield or working harder to get what they need. A vicious circle can begin as a result.

When CPRs are degraded, the poor suffer, particularly the women and children who have to spend more time in gathering fuelwood and fodder. A poor family in a lush green, clean and abundant environment is not as deprived as it would be with the same income in a degraded environment. When fish are plentiful, there is no reason to remain hungry. Take away the benefits of environmental amenities and the poor become destitute.

With regards tribal or indigenous people, they sometimes have very little or no income from wage employment. M. Keating (1993) estimates that such indigenous people comprise about 40 percent of the population in developing countries like Guatemala, Peru and Bolivia, 12 percent in Mexico, 4 percent in Canada and Australia, 7 percent in India, 9 percent in the Philippines and 1 percent in Malaysia. Much of their income may be from natural environmental resources. When development disturbs these natural resources, their lifestyles are threatened. In fact the incidence of poverty is higher in the tribal population than in the rest of the society. Economic development, expansion of cultivated areas and commercial logging have reduced the size of their forests. Similarly, in Brazil, large areas of forests were cut down to promote cattle ranches. Often indigenous people are excluded from nature parks or ecological reserves. This is a mistake as these are the people who in fact have saved these parks in the past and can do so for the future.

Even when the natural resources on which the poor depend are maintained and they are able to get ample fuel, fodder, fish, etc. from them, the poor can still be victims of air and water pollution. Smith (1987, 1993) has extensively documented the exposure poor women and children get from the use of bio-fuels, and the resulting indoor air pollution. This has serious consequences for their health and well being. It is estimated (see Table 12-4) that globally, in 1996, 2.2 million persons died prematurely due to indoor air pollution and 0.51 million due to outdoor air pollution. Provision of cleaner fuels on more efficient stoves could help in both improving

Table 12-4. Air pollution takes its toll - regional picture in 1996 (units: thousand persons).

Regions	Deaths from Indoor Pollution in Thousands		Deaths from Outdoor Pollution in Thousands	Total
	Rural	Urban	Urban	
India	496	93	84	673
Sub-Saharan Africa	490	32		522
China	320	53	70	443
Other Asian Countries	363	40	40	443
Latin America and the Caribbean	180	113	113	406
Industrial Countries		32	147	179
Arab States			57	57
TOTAL:	1,849	363	511	2,723

Source: World Health Organization (1997).

human condition and in reducing pressure on common property resources.

Unlike with most other fuels, air pollutants due to biofuels are largely released directly into homes where the people are inside or near their houses at mealtimes every day. Therefore, the amount of health damage per unit of emissions caused by pollution released indoors can be more than a thousand times greater than that of a smokestack outside of town. Thus, although these emissions are relatively modest, the actual exposure of people is significant in many millions of households around the world.

Several health effects are suspected: respiratory infections in children, chronic lung diseases, lung cancer in adults, as well as adverse pregnancy conditions such as low birth weight and stillbirth, in cases of women exposed during pregnancy.

Acute respiratory infections (ARI) such as pneumonia are the chief cause of death for children under 5 years old and the largest single cause of lost years of life in the world as a whole. Together, the complex of ill-health comprised of the inter-related problems of ARI, diarrhea, and children's diseases such as measles, are the main reason why developing-country children under 5 account for about half the total burden of ill-health in the world, although only making up about 10 percent of the population (WHO, 1997)

Table 12-5. Environment and DALYs.

Disease	DALYs (Millions per year)	Relevant Environmental problem	Percentage of DALYs that would be averted by Feasible Environmental Interventions	Preventive Strategies
Acute Respiratory	119	indoor air	15	-improve ventilation -improve cookstoves -provide electricity to rural households and urban poor.
Diarrhea	99	sanitation, water supply, hygiene	40	-improve quality of drinking water used by improving accessibility and reliability of water supply. -improve sanitation improve hygiene (behavior changes include washing hands, boiling water, preventing casual use of un-protected sources).
Intestinal Worms	18	sanitation, water supply, hygiene	40	-same as for diarrhea -reduce need for contact with infected water.
Tropical Cluster (includes schistosomiasis, typanosolmiasis, and filariasis).	18	sanitation, garbage disposal, vector breeding around the home	30	-reduce need for contact with infected water. -control snail population. -filer water
Tuberculosis	46	crowding	10	-improve housing quality and quantity.
Chronic Respiratory Diseases.	41	indoor air pollution	15	-same as for acute respiratory infections.

Sources: WRI (1998), World Resources Report 1998 - 1999.

Note: WRI defines DALYs as follows. "DALYs combine losses from premature death (defined as the difference between the actual age of death and life expectancy at that age in a low-mortality population), and loss of healthy life resulting from disability. In simple terms, a DALY strives to tally the complete burden that a particular disease exacts. Key elements to consider include the age at which disease or disability occurs, how long its effects linger, and its impact on quality of life. (Losing one's sight at age 7, for instance, is a greater loss than losing one's sight at 67. Similarly, a bout of acute illness that is over quickly counts less in the DALY calculation than one that leaves lingering weakness, such as persistent worm infections.)"

Table 12-6. Consumption disparities between developed and developing countries.

Category	Products	Ratio of Per Capita Consumption	
		Developed / Developing	USA / India
Food	Cereals Milk Meat	3 8 6	6 4 52
Forest	Round Wood Sawn Wood Paper etc.	1 11 14	6 18 115
Industrial	Fertilizers Cement Cotton & Wood Fabrics	5 3 3	6 7 6.4
Metals	Copper Iron & Steel Aluminum	19 13 19	245 22 85
Chemicals	Inorganic Chemicals Organic Chemicals	20 17	54 28
Transport Vehicles	Cars Commercial Vehicles	24 125	320 102
Fuel & Electricity	Solid Liquid Gas Electricity Total Energy	6 10 19 13 10	14 61 227 46 35
Global CO_2 Emissions	Total Emissions Cumulative (1950-88)	8	27

Source: Parikh Jyoti et al (1991), "Consumption Patterns: The Driving Force of Environmental Stress", Indira Gandhi Institute of Development Research, prepared for the UNCED secretariat.

The problem of air pollution is also severe in many urban areas. Air in most cities of developing countries is highly polluted due to industrial growth and increase in the use of motorized transportation.

This report reveals that, for cities where data is available, in 11 large cities of China with a population of 66 million persons and 12 large cities in India with a population of 66 million persons, the level of total suspended particulates (TSP) exceeded the World Health Organization (WHO) prescribed guidelines. (Note: the data relates to different years from 1991 to 1995). Moreover, in all 11 Chinese cities and 9 of the Indian cities as well as in Manila, Bangkok and Mexico City, the TSP level was more than double the WHO guideline.

Urban slums are often located near factories, by roadsides exposed to noise and traffic fumes, near open sewer lines and dumpsites and on swampy lands. It is only on such unwanted sites, that the poor are permitted to squat. At such sites pollution concentration is likely to be higher. Thus the poor are much more exposed to urban air pollution than what the data even suggests. For example, in the Bhopal gas tragedy, the poor slum dwellers that lived next to the Union Carbide factory suffered most.

As already mentioned urban air pollution is estimated to have caused some half a million deaths in 1996. This is mostly in developing countries and among these, most of the victims are likely to be the poor. Improving urban air quality can relieve the poor of this burden of disease and death.

The impact of water pollution on the poor is equally, if not more, severe. As many as 1.4 billion persons in developing countries lack access to safe water (UNICEF, 1997) and 2.9 billion have no access to adequate sanitation (UNDP, 1997a). The traditional environmental health problems arising from insufficient clean water and sanitation are compounded by industrial and agricultural chemical pollution. Along with feces, pesticides now also pollute the drinking water for many. Many chemicals are found in food and water. These include lead, aflatoxins, nitrates, halogenated organic chemicals, aluminum, arsenic and mercury.

These conditions impose their own toll on the poor. WHO estimates 17 million deaths a year from infectious diseases. What is important to note is that with feasible environmental action significant reduction in the deaths and diseases is possible. This fact is clearly illustrated when one considers Disability-Adjusted Life-Years (See Table 12-5).

All of the above call for a drastic change in development patterns. The need for a change becomes even more obvious when we consider the global environmental implications, the main one being that of climate change. This threat of climate change, however, is mainly the result of the consumption patterns of the rich.

The developed countries have only 24 percent of the world population, but their share in global consumption of various commodities ranges from 50 to 90 percent. (See Table 12-6.) Even for the products that fulfill basic needs like cereals, milk and meat, the consumption of the developed countries constitute 48 to 72 percent. The consumption shares (in percent) of the developed countries for other products are: 60 for fertilizer, 81 for paper, 86 for copper and aluminum, 80 for iron and steel, 85 for chemicals and 92 for cars. This also means that their per capita consumption is as high as 3 to 8 times that by the developing countries for items of basic needs and 20 times and more for items like chemicals and vehicles. The share of global energy consumption by the developed world is 75 percent. Obviously, environmental stress is proportional to these consumption levels. The developed world is responsible for about 70 percent of the annual total carbon dioxide (CO_2) emissions and 77 percent of cumulative emissions over 1950-88. The per capita carbon emissions from the rich countries are 7 times those of the developing countries.

It is easy to see how unsustainable this situation really is. We have seen that globalization and the communication revolution have turned this earth into a global village. Everyone aspires to the best life style possible. Suppose the world's population stabilizes at 10 billion persons and everyone has the same consumption pattern as the average U.S. person. Carbon emissions would then exceed 55 billion tonnes per year! Even the most skeptical non-believer in climate change would not be comfortable with this level of carbon emissions.

4. NEW DEVELOPMENT PATTERNS

The present development patterns with the forces they generate and their consequences seem to lead to an unsustainable situation both environmentally and socio-politically. These trends are environmentally unsustainable, both at the local as well as at the global level. When local environmental degradation and its adverse impacts are accounted for, growth seems somewhat illusory. This calls for a change to a much more environmentally sensitive pattern of development. If environment is to be preserved, it will not be done merely by economic instruments. It will also require socio-political change in which people are empowered with rights that motivate them to preserve the environment.

While one may argue that the changes needed to meet the challenge of local environmental problems do not call for a change in the fundamental characteristic of present development patterns, the changes required to meet the threat of climate change are much more fundamental.

We will have to find new technologies that can produce more and more goods with less emissions. This could be done by relying on new energy saving technologies, or switching to renewable energy sources or a non-carbon emitting technology such as nuclear.

Some possible scenarios have been painted. For example, one by the International Institute for Applied Systems Analysis (IIASA) relied on large nuclear stations situated on remote islands and a hydrogen fuel based economy. Our acceptance of nuclear power as an option has gone down considerably since that scenario was presented and many people will be very uneasy with such a future. On the other hand energy conservation and renewable scenarios have also been shown to be feasible. The Zero Emissions Research Initiative (ZERI) of the United Nations University Institute of Advanced Studies is a further elaboration of such ideas.

These alternatives rely on new technologies. If appropriate economic incentives are given, one might expect that the needed technologies will be developed. Yet two different questions arise. First, what kind of incentives are needed? Second, can we institute a system with such incentives that may lead to a globally sustainable outcome? For example, would we institute, in time, a system of tradable emission quotas which are allocated in a just and fair way e.g., on a per capita basis? If not, the countries which develop these new technologies would command substantial monopoly power. With fairly allocated tradable quotas, technology would be priced competitively in the market. Without them, we might see an even more unequal world than the one we currently have.

The growing inequity, which would be aggravated further by this technological scenario could lead to a different kind of unsustainability. The U.S. has a great advantage in developing new technologies. It is a society which encourages individuality. It has an educational system that is superb and widespread. It has a large pool of personnel in fields of science and technology. It has a system which provides research support on merit. The federal government and U.S. industry spend a lot of money on research and development. The U.S. also attracts from all over the world the brightest and the best of those with an entrepreneurial spirit. Finally it has a unique system of venture capitalists. All these have produced many innovator-millionaires, and some billionaires, a record unmatched by any other country.

Even though the U.S. has this advantage, or perhaps because of it, there is a much greater drive now to privatize knowledge and technology through regimes of intellectual property rights. How this regime is set up can have a profound impact on development patterns.

A technology-driven scenario could lead to growing dominance by a few countries and firms. Such a development pattern may not be socio-politically sustainable. Any dominance is resented by the dominated; and when the dominance is attained fairly - as is the U.S. technological dominance - it becomes even more galling.

Growing fundamentalism in the world may be a sign of resentment against such dominance. Any increased dominance might lead to increased fundamentalism. Unfortunately in our global village of instant communication, hatred feeds on itself and spreads faster than understanding. By the time one is confronted with the ugliness of hatred, it has infected many more people.

5. CONCLUDING COMMENTS

Thus in thinking about sustainable development patterns, we should consider a holistic scenario which goes beyond technology. We should analyze the institutions, social and political that can lead to a sustainable world.

If we recognize that the fundamental driving force behind unsustainability is the consumption pattern, we can think of an alternative. We can think of a development pattern in which the desire and demand for material consumption is reduced. After all, we recognize that human happiness, well-being, and fulfillment do not solely depend upon material consumption.

One can conceive of alternative lifestyles in which one is happy with less. All great religions have preached this, but mankind is moving away from it. Can we think of ways in which such scenarios may be made more likely? What kind of development patterns and social political structures could lead to that ideal, so eloquently expressed with remarkable brevity in a Hindu philosophical book, Ishavakya Upanishada, "Ten Tyakten Bhunjithaha" which means "enjoy it by giving it up".

Chapter 13

NEW DEVELOPMENT PATTERNS IN ASIA - Securing Economic Growth, Social Justice, Healthy Environment, Sustainable Human Development and Good Governance in the 21st Century

RYOKICHI HIRONO
Seikei University, Tokyo

1. INTRODUCTION

Since the early 1980s the governments of most countries in Asia, under increasing pressure caused by economic globalisation, focused their national priorities on ensuring macroeconomic stability and growth by accelerating economic reforms, including deregulation, shifting to free market economies, and trade/investment liberalisation. They did this far more aggressively than their previous governments had done during the 1960s and 1970s. Major fiscal, financial and corporate reforms, together with public administration and political reforms, have been introduced to varying degrees, with varying success. Economic growth and globalisation, however, have not succeeded in restraining the on-going deterioration of conditions such as income distribution, poverty, environment and human development issues and governance.

Deeply concerned with the adverse impacts of these macro- and microeconomic reforms on the weaker segments of society, industry and regions, most governments in Asia have begun to explore policies that would ensure the benefits of economic growth distributed to all segments of society and all regions, minimize adverse social and environmental impacts and promote good governance. It is now hoped that these Asian governments, by so doing, will go far beyond the prevailing development policies that still tend to give the top priority to sustained economic growth and that there will be positive effects for equity of distribution, both with current and future generations. It is also hoped that they will seriously begin for the

K. Matsushita (ed.), Environment in the 21st Century and New Development Patterns, 291–300.

twenty-first century to reorient the current growth-bound policies toward new development patterns that ensure economic growth consistent with social justice, environment, sustainable human development and good governance. These goals may be achieved through major policy and institutional reforms and accelerated investment in human capital, science and technology.

This paper is intended to stimulate debate on alternative approaches to development, not only in Asia, but also in the entire world.

2. LONG-TERM VISION FOR ASIAN COUNTRIES FOR THE 21ST CENTURY: GREAT CHALLENGES AND OPPORTUNITIES

2.1 Globalization, Sustained Economic Growth and Economic Crises

Since the mid-1960s many fundamental policy and institutional reforms have been introduced in economic and social sectors by successive administrations in Asian countries, producing a reasonable level of macroeconomic stability and high rates of sustained economic growth and laying the essential foundations for sustained growth in the twenty-first century. The steady international pressure for economic, political and social globalization added momentum to these domestic reforms (which varied in degree among countries), which promoted further integration into the world economy (through export-oriented industrialization and liberalization of foreign exchange, trade and investment). One outcome was the financial and economic crises in many countries of the region at the end of 1997. In some cases, the crises even stimulated political crises and adversely affected the economic and social well-being of the population, particularly the weaker segments and regions.

2.2 Impact on Social Dimensions of Development

While steady progress has been made in most Asian countries in terms of industrialization and economic restructuring as well as some dimensions of sustainable human development, such as life expectancy, literacy and per capita income, many issues have not been resolved. In some cases conditions have actually deteriorated in a number of social and environmental dimensions of development, such as unemployment and underemployment, income disparities both in rural and urban areas, poverty, air, water, soil and noise pollution and crime. In many countries problems of governance have also remained unresolved. The recent Asian financial and economic crises worsened the socio-economic situations in most countries of East Asia, with the possibility of spillover effects in South Asian countries.

The middle class has grown rapidly through three decades of sustained economic growth. It is demanding higher standards of living, better health, education, job opportunities and environment, and good governance. It expects political freedom, the respect of human rights and participatory decision-making, and has the strong support of the international community, in particular from international non-governmental organizations (NGOs) in these expectations. Most governments in the Asian region will continue in the coming decades to feel growing political pressure to meet their people's expectations despite enormous economic, social, technological, administrative and political constraints, both in the short- and long-run. Difficulties will be especially serious in tradition-bound developing countries such as in South Asia and countries under the communist rule.

2.3 Need for National Consensus on Long-term Visions in Order to Attract Stable International Assistance

In this context, it is all the more important that governments of Asian countries at the earliest possible time spell out their long-term visions for their respective populations that are both desirable and realistic, covering at least the first quarter of the twenty-first century. As much as possible, it is also essential that these visions be based on consensus of all the segments of the population and their representatives at the national and local levels. The visions should tell what kind of society the people want to achieve in terms of political, economic, social and environmental objectives and map out how they can be achieved in terms of priorities, depth, timing and sequences, taking into account all resources and political constraints. In the increasingly globalized world these visions should also be acceptable to the international community, reflecting the common values, standards and rules now and in the future, ingrained in a large number of international agreements, protocols and conventions.

It is likely that the public of these Asian countries will call for efficient, just, fair and dynamic societies which protect political freedom and human rights, sustainable and reasonable rates of economic and income growth and employment opportunities, fairer distribution of income and poverty reduction, and environmental protection. However, it is not enough to just state the desires. To realize such societies by the end of the first quarter of the twenty-first century broad-based strategies, specific targets and concrete policy measures are necessary. Where possible these should be in quantitative terms, based upon long-term projections of resource requirements. Consideration is needed about whether these requirements can be met domestically, and whether foreign resources are also needed.

2.4 Challenges Facing Asian Countries

The greatest challenge facing Asian governments today in presenting such long-term visions lies not in any shortage of domestic technical and professional competence in formulating such visions. Rather, the challenge will be the political difficulty in swiftly reaching national consensus on such visions. In addition, challenges will be posed by the lack of domestic capacity to implement such visions, arising from the resurgence of enormous empires of vested interests among industries and people that resulted from sustained economic growth and development, as well as the traditional socio-political structures. Strong political leadership is needed in order to spell out long-term visions, which can attract broad citizen support including from academia, the private sector, cooperatives and NGOs, and to build capacity both inside and outside the government bureaucracy.

3. MUTUALLY REINFORCING REFORMS OF ECONOMIC, SOCIAL, ENVIRONMENTAL AND GOVERNANCE POLICIES

3.1 Need for an Integrated Approach

Previous governments tended to focus on macro- and micro-economic reforms aiming for sustained macroeconomic stability and growth through greater efficiency in resource allocation. However, in the twenty-first century Asian governments should demonstrate integrated approaches toward reforms of economic, social, environmental and governance policy, while making clear the time horizons required to achieve the reforms, as well as the policy priorities. These approaches should integrate social justice, the environment, sustainable human development and - increasingly important - good governance.

There are two key elements to consider in these integrated approaches. The first is the urgent need for political responses as economic reforms and growth impose increasingly negative impacts on society and the environment, as members of the national and local legislatures have repeatedly expressed. The impacts are particularly noticeable on the weaker members of society as well as on the natural and social environments. The second is the enormously positive contribution of investments in the social sector (i.e., in human and institutional capital) toward sustained economic growth and expanded incomes in market economies, and toward good governance.

3.2 Growing Recognition for Policy and Institutional Reforms for Enhanced Social Justice, Environment, Sustainable Human Development and Good Governance

Despite substantial economic benefits of macro- and micro-economic reforms and sustained economic growth, which have occurred since the mid-1960s, Asians generally have not enjoyed the full benefits of political freedom, human rights and good governance. They have also been experiencing the adverse impacts of these reforms on their daily lives in terms of high unemployment levels in urban areas and underemployment in rural areas, as well as poverty, environmental deterioration and growing crime rates. These issues of good governance and social and environmental deterioration, which are worsening in the shadow economic affluence, should be urgently addressed by governments in the region. Recognition is actually growing within these governments that macro- and micro-economic reforms cause damage if insufficient attention is paid to political, social and environmental policy and institutional reforms. Successful macro- and micro-economic reforms require reforms of political, social and environmental policy as well as institutional reforms, and the reverse is also true. Stated another way, sustained economic growth cannot be achieved without social justice, environment, sustainable human development and good governance. Many Asian countries have started these reforms, but they happen to coincide with major policy shifts alluded to in various recent statements by the managing director of the International Monetary Fund and president of the World Bank.

3.3 Traditional Versus New Concepts of Governance, Social and Environmental Policy

In almost all Asian countries that have mixed economies it has been generally accepted that governments should be responsible for maintaining political and economic security, promoting economic and social development, and enhancing social and environmental welfare. Exceptions include China, the Democratic People's Republic of Korea (DPRK), Laos and Vietnam, which have been under socialist regimes for many years. In some Asian countries national interests have taken priority over the interests of the people, who have been taught that their duty is to serve their country. Issues or opposing views about governance were almost nonexistent, regardless of how inappropriate the management or methods of elections; generally it was assumed that the government installed by the party winning the general election held political legitimacy. Because programmes relating to social and environmental policy have been narrowly interpreted as targeting the weaker members of society or as environmental relief programmes, social and environmental expenditures have been perceived either as unproductive or a drain on budgets of central and local governments. Thus in most Asian countries governance never became a major political issue, and successive governments until recently gave only a low priority to social and environmental policy. They put their faith into the so-called trickle-down theory and recited the belief that sound economic policy was also the best social and environmental policy.

This thinking in Asian governments needs to change course by 180 degrees. Today the concepts of governance based on traditional assumptions of government, the initiatives of socialist regimes relating social and environmental welfare, and the economic policies of many mixed economies based on the trickle-down theory - all these are no longer acceptable. Asian countries should serve the people with whom the national sovereignty rests; all government policies and practices should be formulated and implemented for the benefit of their citizens. Government policies and practices should be subjected to the scrutiny of the representatives of the people, the national parliament and local assemblies, with active participation by the public and civil society organizations. No reason should be accepted to bypass national parliaments and local assemblies (including public hearings) on the grounds that their procedures would dilute the effectiveness of policies, distort programmes, delay action or reduce efficiency. Programmes relating to social and environmental policy should be considered as fundamental and crucial rather than as merely remedial, and related expenditures as investments rather than simply as expenses. They are essentially investments in human and institutional capital which earn the highest possible return on investment in the long run, as proven in all industrial and some advanced developing countries. Social and environmental programmes should be integrated with economic programmes and given the top priority by Asian countries since they are the most productive investments, if governments share the same aspirations as their citizens and are truly concerned about improving economic performance through sustained economic reforms, while arresting further deterioration of social justice, the environment and sustainable human development.

While the social and environmental programmes integrated into the economic programmes could yield the highest return on investment, the existing ones in many Asian countries have a number of weaknesses both in terms of policy, implementation and financing, as pointed out by many reports of international organizations. To improve them in terms of quality and cost effectiveness, a thorough and critical review of the entire range of social and environmental policies is urgently needed.

4. DEVELOPING NATIONAL STRATEGIES IN ASIAN COUNTRIES COMPATIBLE WITH NEW DEVELOPMENT PATTERNS

4.1 Economic Policy Frameworks -- Major Weaknesses and Reforms

A tendency exists for inefficient resource allocation at the macro and micro levels as a result of too many governmental regulations and interference. Governments should limit their major scope of activities to the formulation of national development strategies and programmes in line with the long-term visions discussed above, and to the adequate provision of economic and social infrastructure, including education and health. In formulating and implementing national development strategies and programmes, governments should use all the policy instruments available to them, including fiscal and financial incentives, to promote economic growth and restructuring. They should also facilitate research and development to assist sunrise industries, to improve productivity in all sectors of the national economy in order to be competitive in the international market and to provide workers in declining industries with social safety net. Governments should strive to maintain price and economic stability, restore fiscal and external balances, reduce poverty and foster scientific and technological innovations that save energy and resources and are environmentally friendly. All environmental costs arising from market transactions should be taken into account in product prices through taxes and regulations, so that market prices reflect all social costs associated with private sector activities, including the production and distribution of goods and services. These taxes and regulations are to limit environmental deterioration as well as other social costs not mentioned here.

Governments are still heavily involved in the production of goods and provision of services, although many of these can be done much more efficiently if left to the private sector. Governments are encouraged to privatize state-owned enterprises that compete on an unequal footing with the private sector, abolish state monopolies, deregulate market transactions and promote competition. Financial and fiscal policy should encourage improved efficiency in resource allocation at the macro and micro level, and also enable the private sector to meet the changing needs and requirements of the market through technological innovations.

The flaws of lacking transparency and accountability have been recognized in governments when they formulate and implement public investment programmes. Examples include (1) bureaucratic rigidity when sticking to old development programmes and projects that have lost their usefulness under changing economic and social conditions, (2) political interference in investment programmes, and (3) allowing contractors opportunities to overprice and provide substandard quality, by not using competitive bidding. These ways of handling affairs are to the detriment of not only taxpayers but also to consumers and users of investment projects such as highways, schools, hospitals and airports. Governments should constantly review their public investment programmes so as to tailor them to the changing needs and requirements of national and local development plans and programmes consistent with maintenance of competitive strength of their economies, minimize regional and urban-rural disparities and provide long-term economic and social security for citizens. Governments should also do everything possible to eradicate these shortcomings by introducing competition at all stages of public investment programmes and project cycles, including the formulation, design, implementation, monitoring and evaluation stages. The same attention is needed for recurring government procurement programmes.

4.2 Social Policy Frameworks -- Major Weaknesses and Reforms

It could be said that all Asian countries lack integrated and consistent social policy frameworks to improve the economic and social wellbeing of both the productive and nonproductive members of the population. Correcting this shortcoming should become a top priority.

In most Asian countries, macro- and micro-economic reforms have so far produced substantial economic benefits, and in socialist countries have facilitated the transition from centrally planned to market-oriented economies. In the future, appropriate pricing policies should be put in place to internalize the social costs of these reforms. Of course, the reforms should not be abandoned just because of the shortcomings in these policies.

Until now, social policies to provide a social safety net for those segments of the population that require social assistance and protection have been inward looking and passive. In the future, as a form of investment in human and institutional capital, social policies should be more proactive.

Developments in the social policy sector, such as in education, training and health services, have emphasized the amount of resources going in, rather than the results coming out. This tendency should be reversed in the future.

Many examples exist of the misguided and wasteful use of public resources at the macro level. Outdated national and local regulations lead to a loss of flexibility in resource utilization. Bureaucrats classify sectors of society that are inconvenient to them in a way that is easy for them to manage in their programmes. However social security and services should instead be suited to the needs of the recipients rather than the needs of the administrators.

A lack of decentralization in the formulation and management of social policies results not only in delays in budget transfers from national governments and agencies to local governments, but more importantly in the lack of an innovative spirit and competence in the organizations and people at the local level.

The lack of organizational and individual incentives has been quite conspicuous at the micro level, not only increasing the risk of not meeting the quality and time objectives of social policy project activities, but also reducing their cost-effectiveness. Administrators and recipients of social services should have a scope of options that allows them to create the maximum benefits.

The mobilization of private sector enterprises, cooperatives, non-governmental organizations (NGOs) and community-based organizations (CBOs) has been inadequate. This has not only increased the fiscal burden on the national and local governments but more importantly has reduced the benefits and cost-effectiveness of social policy programmes. All development entities in a country should be able to participate in the formulation and implementation of social policies and programme planning. The role of government should be limited to preparation of the basic management structure of programmes, and anything else should be left for non-governmental development organizations.

As a result of inadequate coordination among national government ministries, departments and agencies, aid funds from overseas have not been used effectively, leading to duplication, fragmentation and inconsistencies in programmes. Coordination of the formulation and management of social policies should be implemented with the involvement of both central and local governments, but the coordination of funds from overseas should be done by the central government.

Introduced below are some new approaches regarding concepts about social policies:

(a) The authority of central governments should be limited to proposing and supervising new social policy frameworks and programmes. Regarding financing, governments should limit themselves to the basic elements (basic healthcare, basic education and social assistance/protection).

(b) Multiple-track approaches (i.e., involve public, private, cooperatives, NGO and local community sectors) should be introduced wherever possible, in order to involve all sectors of society in implementing and cost-sharing of the new social policy frameworks and programmes including the basic elements.

(c) A principle of independent budgeting should be strictly adhered to, and government resource allocation for social insurance programmes, including health insurance, unemployment insurance, workmen's compensation and pensions, should be kept at the lowest possible levels. In other words, allocations for social insurance should be kept at the lowest possible levels, and any other requirements should be entrusted to a private insurance system.

(d) A variety of cost recovery mechanisms should be introduced in education, particularly at the tertiary level.

(e) Tax and other incentives should be provided to private businesses, cooperatives, NGOs and CBOs in order to enhance the quality of the social policy programmes, particularly for health care, human resources development and employment.

(f) Decentralization of functions to local governments should be promoted for implementing and monitoring not only social assistance programmes but also other overall components of social policies and programmes.

Afterword

Kazuo Matsushita
Institute for Global Environmental Strategies

This book is the second of a series by IGES.

The individuals involved its preparation would be delighted if it stimulates thinking about the future of the environment and new forms of development, as humanity stands at the dawn of the twenty-first century.

The era since the Industrial Revolution was characterized by socio-economic systems which stimulated dramatic increases in the amounts of material resources and energy used. They also drove mass production, mass consumption, and mass disposal–these were the faces of development, especially in the twentieth-century. It should be noted that this type of industrial activity and social system did not spread equally worldwide. It was concentrated in advanced, industrialized countries, particularly in North America, Western Europe, and Japan. Many people have pointed out that this type of development pattern is NOT sustainable. However, the current reality for developing countries struggling with poverty compels them to seek tangible development models from today's developed countries, of which America's model appears to be at the apex. In addition, the globalization of economies, which is racing ahead at breakneck speed, is escalating problems at the global scale. This overall situation is the reason new types of development patterns are needed for the twenty-first century.

This book presented a number of issues that must be considered when discussing these new development patterns. For readers, perhaps the book has catalyzed a new awareness of various issues. Perhaps you find yourself contemplating questions such as those below.

How can equity and justice be realized on a global scale? And what concepts of development can respond to these types of challenges? How can we use a strength of globalization–the fact that it promotes local autonomy and decentralization, and this emphasizes diversity and the uniqueness of local cultures?

What kind of societies will new technological innovations lead to–innovations such as those we see in the information technology (IT) revolution? Will they contribute to the formation of sustainable societies? Will each person, including

K. Matsushita (ed.), Environment in the 21st Century and New Development Patterns, 301–303.

those living in developing countries, have opportunities to utilize this information and more chances to build relationships? Or will new types of discrimination and differences appear between people?

What is the potential for developments in renewable energy and improved energy efficiency? What about advances in material efficiency, recycling technologies, and technologies to eliminate toxic substances? How can we expand the availability of these technologies and social systems? How can we ensure equity and social welfare on a global scale?

How should frameworks and roadmaps toward societies based on sound material cycles be made more tangible? Can developing countries themselves–in Asia and other regions–present new models to the world? What lessons can we learn from each other?

How can we realize change in the foundations of our national governments–such as the introduction of economic measures like a comprehensive environmental tax system? Or innovations and the democratization of environmental policy, like environmental assessments, information disclosure, and guarantees of public participation? Or the integration of environmental and economic policies?

We think that we at IGES must take up these issues with readers. We welcome comments, ideas and suggestions about new development patterns project and issues raised in this book (IGES web-site: http://www.iges.or.jp; please give comments to < k.matsushita@iges.or.jp >). In the year 2002, a conference is to be held at the United Nations level in order to conduct a comprehensive review of the implementation of Agenda 21. It is being called the Rio Plus Ten conference, marking the ten years that will have passed since the Earth Summit in Rio de Janeiro. As one target on the road to that landmark event, countries must move toward realization of a society based on sound material cycles, and contribute to regional cooperation and the formation of international frameworks.

The production of this book has its origins in early discussions of the IGES New Development Patterns Project, which was launched in the middle of 1998. The first steps were to organize a workshop to sort out the key issues, and to have experts from inside and outside Japan write issue papers. Senior Research Fellow Glen Paoletto played a crucial and active role in the selection of topics and writers, requests for papers, and consideration of the overall concepts. Study groups and workshops were held with some of the authors. I would like to express my heartfelt gratitude to the authors of the papers for taking the time to write the precious manuscripts, despite their busy schedules.

I was responsible for coordinating the editing of this English edition. Randal Helten conducted the detailed checking and editing of the text, with the support of Naoko Miyazaki. A considerable amount of time and effort was needed to verify the details of some of the papers translated from Japanese. In addition, we received valuable advice from Prof. Akio Morishima, Chair of the Board of Directors of IGES, relating not only to the preparation of this book but also to the entire New

Development Patterns Project. Project members also gave useful suggestions. The administrative staff of IGES provided overall support.

Kluwer Academic Publishers provided outstanding assistance in the publication work. I would like to take this opportunity to express my sincere appreciation to everyone who contributed to the publication of this book.

July 2000

MIX
Papier aus verantwortungsvollen Quellen
Paper from responsible sources
FSC® C105338

If you have any concerns about our products,
you can contact us on
ProductSafety@springernature.com

In case Publisher is established outside the EU,
the EU authorized representative is:
Springer Nature Customer Service Center GmbH
Europaplatz 3, 69115 Heidelberg, Germany

Printed by Libri Plureos GmbH
in Hamburg, Germany